The Many Uses of

Orthogonal Functions

D. James Ben…

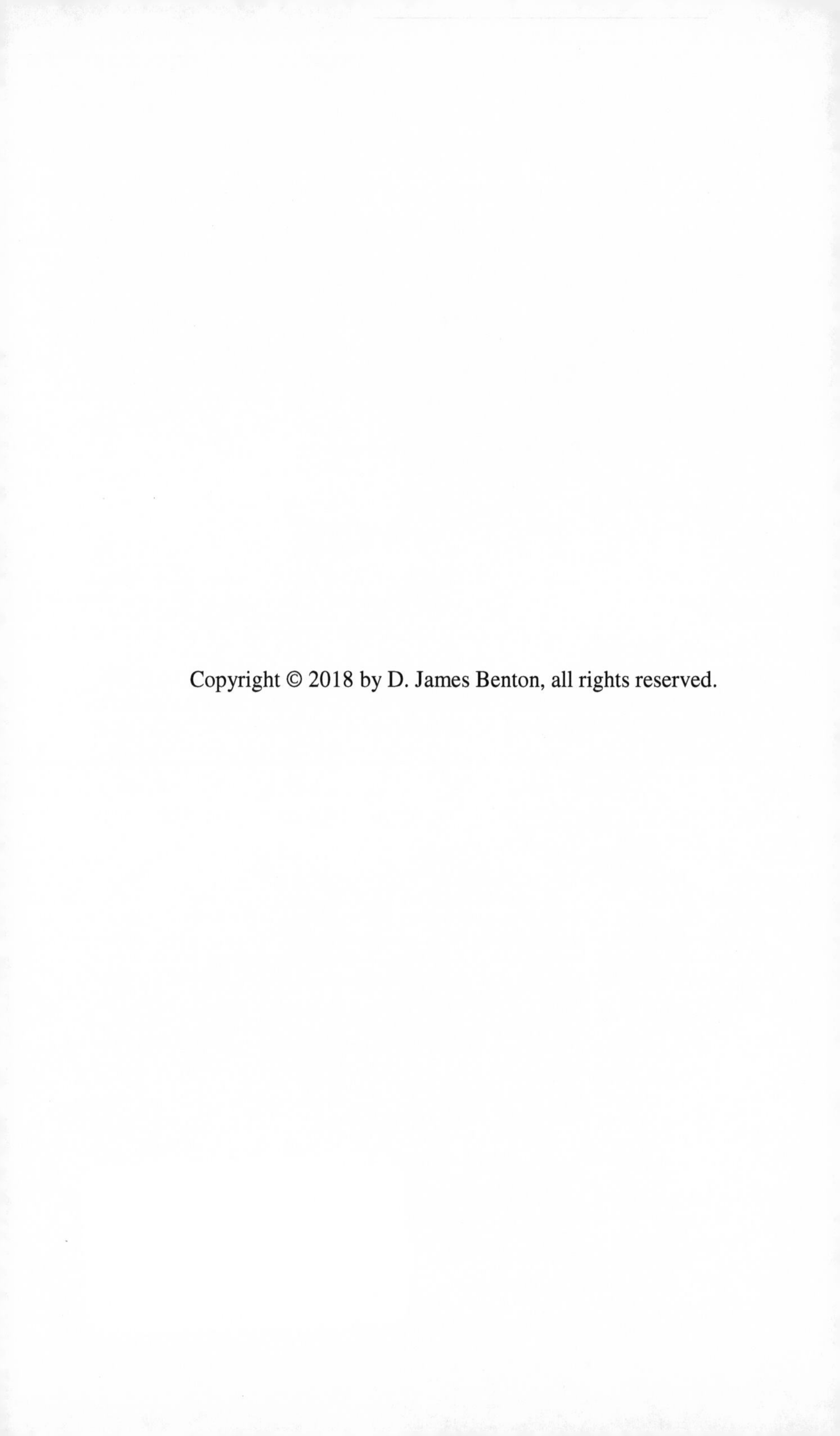

Foreword

Orthogonal functions are clever tools that unlock many mathematical puzzles. Once you've seen this done and understand how they work, you will find many more useful applications. This remarkable area of applied mathematics supports a wide range of technologies, ranging from CAT scans to satellite pictures from space to unraveling the sounds of the deep. Join me on a tour of this fascinating topic in which we will explore data sampling and approximation in both temporal and spatial dimensions.

All of the examples contained in this book,
(as well as a lot of free programs) are available at...

http://www.dudleybenton.altervista.org/software/index.html

Programming

Most of the examples in this book are implemented in the C programming language. A few are implemented in VBA® (Visual BASIC for Applications, or what Microsoft® calls the language of Excel® macros). BASIC stands for Beginner's All-Purpose Symbolic Instruction Code. If you're still using some form of BASIC and haven't yet graduated to a professional programming language, now is the time to do so and there is nothing better than C. It's in a *class* by itself.

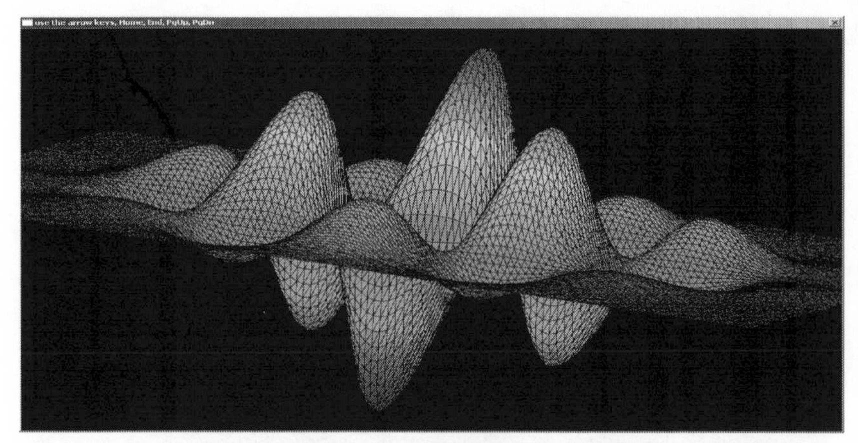

examples\surface\surface.exe Hermite.tb2

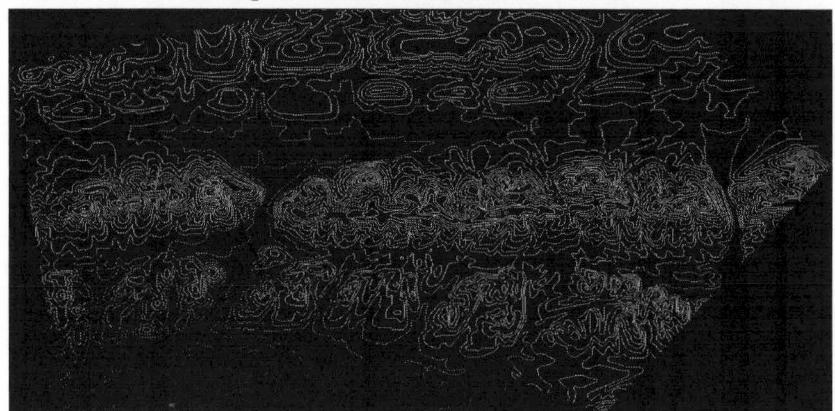

examples\topography\transformed.p3d

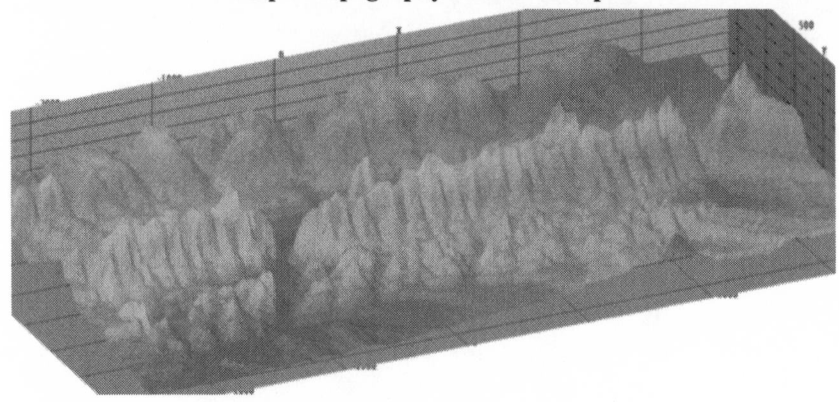

examples\topography\inverse_distance.tb2

Table of Contents

page

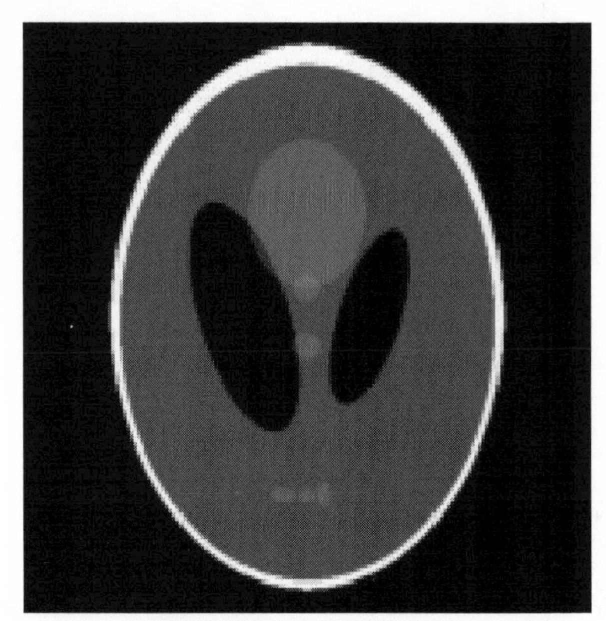

Shepp-Logan Phantom

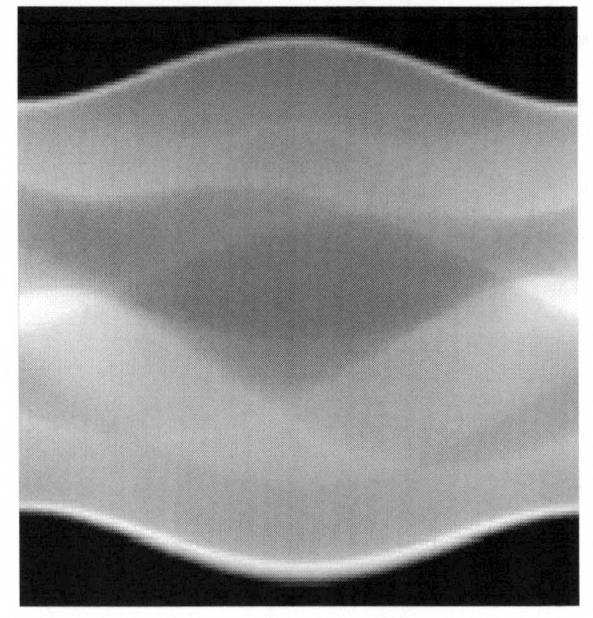

Corresponding Sinogram

Chapter 1. Orthogonality: What Does it Mean?

You *must* see this to appreciate it. While this may seem like a strange place to start, trust me, it will be eye opening... Let's say you want to approximate some data using a simple polynomial of the form $y=c_0+C_1x+c_2x^2+c_3x^3+\ldots$ The residual (i.e., error at each point) is given by:

$$r_i = c_0 + c_1 x_i + c_2 x_i^2 + c_3 x_i^3 - y_i \qquad (1.1)$$

In matrix form, R contains the r_i, C contains the c_i, X contains the x_i, and Y contains the y_i. We can write:

$$R = C^T X - Y \qquad (1.2)$$

The sum of the squares of r_i is equal to $R^T R$, where R^T is the transpose of R. Equation 1.2 can be expanded to obtain:

$$R^T R = \left[C^T X - Y\right]^T \left[C^T X - Y\right] \qquad (1.3)$$

Equation 1.3 can be further expanded:

$$R^T R = \left[C^T X\right]^T \left[C^T X\right] - 2\left[C^T X\right]^T Y + Y^T Y \qquad (1.4)$$

To find the coefficients resulting in the smallest residual, we take the derivative of Equation 1.4 with respect to C and set this equal to zero.

$$0 = \left[X^T X\right]C - X^T Y \qquad (1.5)$$

Solving Equation 1.5 for C yields:

$$C = \left[X^T X\right]^{-1}\left[X^T Y\right] \qquad (1.6)$$

This is the fundamental equation of linear regression seeking the least squares residual. Let's see how this works for a typical problem with four coefficients and a polynomial order of three. The spreadsheet (discrete.xls) can be found in the online archive in folder examples\discrete. The matrices are:

1	1	1	1
1	2	4	8
1	3	9	27
1	4	16	64
X			

1	1	1	1
1	2	3	4
1	4	9	16
1	8	27	64
X^T			

4	10	30	100
10	30	100	354
30	100	354	1300
100	354	1300	4890
X^TX			

69.00	-104.17	45.00	-5.83
-104.17	161.39	-70.83	9.28
45.00	-70.83	31.50	-4.17
-5.83	9.28	-4.17	0.56
$[X^TX]^{-1}$			

1
1
2
6
Y

10
33
119
447
X^TY

$[X^TX]^{-1}[X^TY]$	P	R
0.000	1	0.00
2.167	1	0.00
-1.500	2	0.00
0.333	6	0.00

Because the number of constants equals the number of points in this case, the residual is zero. This seems straightforward enough and is easily implemented in Excel® with functions TRANSPOSE(), MMULT(), and MINVERSE(), as illustrated in the spreadsheet. You will get the same result using LINEST(). Note that Excel's LINEST function returns the coefficients in reverse order. There's even a shorthand way of specifying the powers of X:

```
=TRANSPOSE(LINEST(U1:U4,B1:B4^{1,2,3},TRUE,FALSE))
```

We can set up a program (condition.c) to step through a range of orders to obtain the following table. There are several definitions of the condition number of a matrix, including the ratio of the largest to smallest singular value. We can estimate this important measure of stability by accumulating the product of the pivots. The larger the condition number, the more unstable the matrix, the more susceptible it is to round-off error, and the more meaningless the inverse.

n	condition
2	1
3	4
4	144
5	82944
6	1.19E+09
7	6.19E+14
8	1.57E+22
9	2.56E+31
10	3.36E+42
11	2.35E+55
12	1.26E+72
13	4.45E+91
14	7.78E+114
15	3.95E+140
16	2.19E+170
17	3.78E+202
18	3.51E+240
19	2.88E+279
20	1.#INF

The condition number blows up at n=20, that is, the value is bigger than the largest double precision floating point number. Needless to say, the results are meaningless long before this. Round off can—and will–eat your lunch with high order regressions.

Orthogonal Functions to the Rescue!

What if... instead of using x, x^2, x^3, etc., we used some very simple polynomials. These will have the same order, but will be cleverly formed so that they have a very special property: discrete orthogonality. These special polynomials can be represented by the following sum and products:

$$K_{nm}(x) = \sum_{i=0}^{m} (-1)^i B(m,i) B(m+i,i) \frac{\Pi(x,i)}{\Pi(n,i)} \quad m = 0,1,2,\ldots,n \tag{1.7}$$

The product in Equation 1.7 is similar to a factorial:

$$\Pi(n,i) = n(n-1)(n-2)\ldots(n-i+1) \tag{1.8}$$

The polynomials for $n=4$ are shown in this next figure:

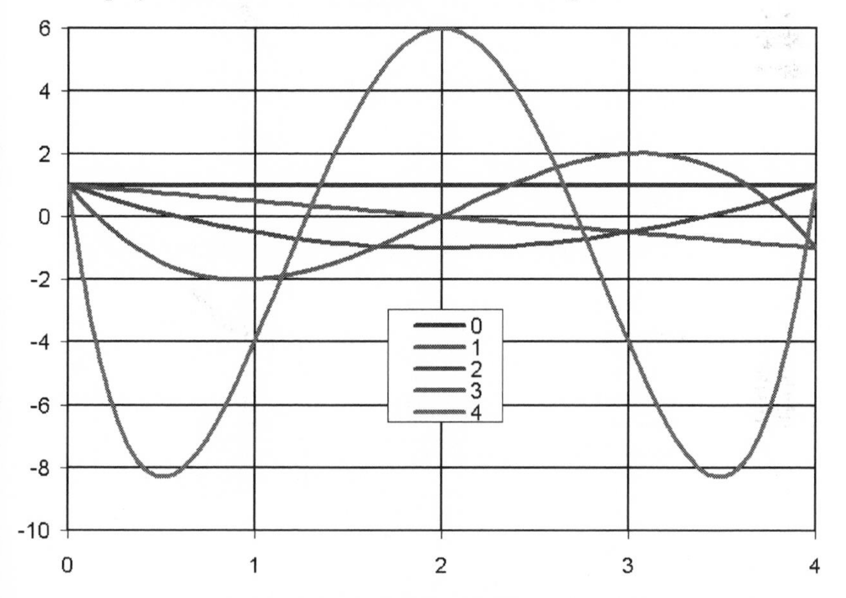

Note that all of the polynomials pass through $y=1$ at $x=0$. These are *orthogonal* over the discrete set of points, x_i, which condition may be stated:

$$\sum_{x=0}^{n} K_{ni}(x) K_{nj}(x) \begin{cases} \neq 0 \; if \; i = j \\ = 0 \; if \; i \neq j \end{cases} \tag{1.9}$$

The first 3 polynomials are:

$$K_{no}(x)=1$$
$$K_{n1}(x)=1-\frac{2x}{n} \quad (1.10)$$
$$K_{n2}(x)=1-\frac{6x}{n}+\frac{6x(x-1)}{n(n-1)}$$

These polynomials were first introduced by the Ukrainian mathematician Mikhail Kravchuk in 1929.[1] Additional details may be found in other references.[2,3,4] The VBA code is listed below and may be found in the examples\ discrete folder in the online archive:

```
Function orthopoly(ByVal n As Integer, ByVal m As
   Integer, ByVal x As Double) As Double
  Dim i As Integer, q As Double
  orthopoly = 1
  q = 1
  For i = 1 To m
    q = -q * x / n
    orthopoly = orthopoly + binomial(m, i) * binomial(m
   + i, i) * q
    x = x - 1
    n = n - 1
  Next i
End Function
```

Let's substitute these polynomials into the same spreadsheet and see what happens…

[X]				[X^T] transpose				[X^T][X]			
1	1	1	1	1	1	1	1	4	0	0	0
1	0.333	-1	-3	1	0.333	-0.333	-1	0	2	0	0
1	-0.333	-1	3	1	-1	-1	1	0	0	4	0
1	-1	1	-1	1	-3	3	-1	0	0	0	20

The off-diagonal elements disappear (i.e., go to zero)! We don't have to invert a matrix when all the off-diagonal elements are zero. The inverse is

[1] Kravchuk, M. "Sur une généralisation des polynomes d'Hermite," Comptes Rendus Mathématique (in French), Vol. 189, pp. 620–622, 1929.

[2] Koornwinder, T. H.; Wong, Roderick, S. C.; Koekoek, Roelof; and Swarttouw, R. F., "Hahn Class: Definitions", in Olver, Frank W. J.; Lozier, Daniel M.; Boisvert, Ronald F.; Clark, Charles W., NIST Handbook of Mathematical Functions, Cambridge University Press, 2010.

[3] Levenshtein, V. I., "Krawtchouk Polynomials and Universal Bounds for Codes and Designs in Hamming Spaces", IEEE Transactions on Information Theory, Vol. 41, No. 5, pp. 1303–1321, 1995.

[4] Wylie, C. R., *Advanced Engineering Mathematics*, 4th. Ed., pp. 157-159, McGraw-Hill, 1975.

simply the reciprocal of the diagonal elements. We don't even need to compute the off-diagonal elements, because they're always zero. The matrix operations are gone and all we have to do is perform a few sums to arrive at the coefficients. If we substituted the expanded polynomials and carried out the algebra, we would obtain the same coefficients as before, using x, x^2, x^3, etc., but there's no point. Here's the same thing in 5x5:

1	1	1	1	1
1	0.5	-0.5	-2	-4
1	0	-1	-0	6
1	-0.5	-0.5	2	-4
1	-1	1	-1	1

[X]

1	1	1	1	1
1	0.5	0	-0.5	-1
1	-0.5	-1	-0.5	1
1	-2	-0	2	-1
1	-4	6	-4	1

$[X^T]$ transpose

5	0	0	0	0
0	3	0	0	0
0	0	4	0	0
0	0	0	10	0
0	0	0	0	70

$[X^T][X]$

The same procedure of stepping through increasing orders of approximation can be implemented using instead these orthogonal polynomials. The code required to calculate the coefficients is trivial and can be found in discrete.c:

```
for(i=0;i<n;i++)
  {
  for(S=k=0;k<n;k++)
    {
    P=polynomial(n,i,k+1);
    B=gamma(k+1);
    C[i]+=P*B;
    S+=P*P;
    }
  C[i]/=S;
  }
```

Round off does take over at order 18, but not because of the matrix inversion process. The approximation is reasonable up until that point, plus it taks a lot less computational effort to accomplish the same thing. This is the *power of perpendicular thinking!*

What's the Point?

Why go through all this trouble just to save a little effort inverting a matrix? To answer the age-old question, "How much is enough?" I was tasked with addressing the following issue... We were performing a contractual acceptance test for a billion dollar combined cycle power plant. It was the last possible day to test. The weather that morning was not conducive to performance testing, but improved later on. When the day was complete, the data collected and analyzed, a serious problem became apparent. Depending on which sections of the data were kept and which were discarded, the outcome ranged from large bonus to even larger penalty.

We couldn't simply wait until the following day and do the whole thing all over again, because that would require stipulating late completion and automatically incurring liquidated damages (i.e., a large penalty payment) for the prime contractor. If we threw out one part of the data, it would mean a large bonus paid by the owner to the equipment manufacturer (combustion gas turbines). If we threw out a different part of the data, it would mean an even larger payment from the equipment manufacturer to the owner for under performance. These potential payments were more than I make in a decade and nobody wants to pay out. I was tasked with analyzing the data objectively and providing an impartial mathematical justification that would stand up in court.

How do you know when you've adequately characterized a trend or sufficiently extracted the useful information from a sequence of data? This is the same question astronomers at NASA had to consider when transmitting images from Voyager back to Earth. They didn't have the transmitting power, speed, or accuracy to send everything. But how much is enough? When compressing a photograph into a JPEG so that it's only 35 kilobytes instead 15 megabytes, how small can you make it without losing too much information. A Fourier transform using orthogonal functions (sin and cos) is used to compress a JPEG.

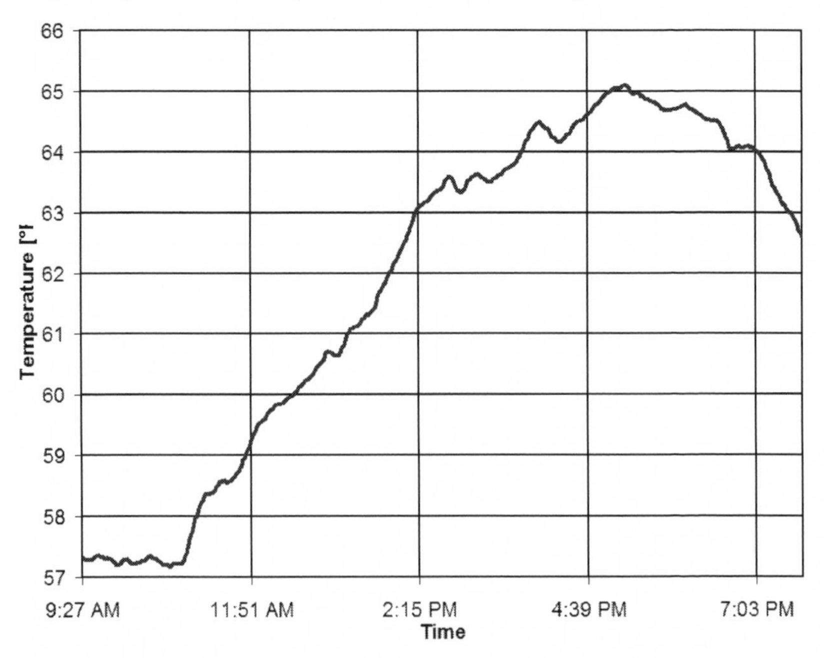

We will use these discrete orthogonal polynomials to analyze the temperature data in the figure above. Recall from the previous figure that all the polynomials all pass through $y=1$ at $x=0$. There were definite criteria for variability and stability of the data, but these depend on the selected time intervals, which were the core of dispute. You will find a program, data file, and spreadsheet in the examples\discrete folder of the online archive with everything needed to perform the analysis. We will consider sequential approximations of increasing order up to 8 terms. The program output is:

```
examples\discrete>data
reading data: data.csv
  617 lines read
  617 data points found
comparing progressive approximations
conventional polynomials
  y=c0+c1*x+c2*x^2+...
  C[3]=55.4187,21.4637,-12.8245
  C[4]=56.8431,4.27342,30.256,-28.7669
  C[5]=57.127,-1.4547,56.1232,-69.092,20.1953
  C[6]=57.4379,-10.9039,122.59,-246.839,220.566,
    -80.2785
  C[7]=57.4608,-11.8865,132.48,-286.546,295.203,
    -146.087,21.9717
  C[8]=57.51,-14.7036,170.83,-500.574,885.591,-
    998.219,638.565,-176.455
orthogonal polynomials
  y=c0*K0+c1*K1+c2*K2+...
  C[3]=61.11,-2.64,-1.15
  C[4]=61.1575,-2.9205,-1.1975,0.2805
  C[5]=61.6,-3.218,-1.66429,0.578,0.0242857
  C[6]=61.4,-3.39714,-1.44643,0.751111,0.00642857,
    0.00603175
  C[7]=61.5786,-3.48107,-1.67679,0.81,0.0751948,
    0.0310714,-0.0169805
  C[8]=61.5325,-3.71417,-1.695,0.999091,0.105682,
    0.0759615,0.0168182,-0.000885781
```

The first (zero order) term in the conventional polynomial approximations are: 55.4187, 56.8431, 57.127, 57.4379, 57.4608, and 57.51. These are converging toward the flat segment between 9:27 and 10:53 AM, which isn't an average and was problematic for one of the parties involved. Note that the other coefficients just keep getting bigger: -12.8245, 30.256, -69.092, -246.839, 295.203, 998.219. This is typical for conventional polynomial regression of increasing order.

The first (zero order) term in the orthogonal polynomial approximations approach the average: 61.11, 61.1575, 61.6, 61.4, 61.5786, and 61.5325. The highest order coefficients keep getting smaller: -1.15, 0.2805, 0.0242857, 0.00603175, -0.0169805, -0.000885781. This is typical of regression using orthogonal functions. The second (order one) term is also approaching a constant: -2.64, -2.9205, -3.218, -3.39714, -3.48107, -3.71417. This is the ramp rate of the data. Even the third (order two) term approaches a constant.

Because all of the orthogonal polynomials pass through unity at $x=0$, we can objectively say that each term of the successive approximation has *captured* the behavior of the data at a level of the magnitude of the coefficients. Once the terms get smaller than the prescribed allowable variability, we stop and evaluate the results. In this particular case there were no bonuses or penalties—an outcome I suspected from the beginning, the plant being neither under or over performing, but as designed.

Best Choice of Methods

As with many other things, the best choice of approaches for any given task is not a simple process of doing it the same way as the last time. Hopefully, you don't use a screwdriver when a wrench or hammer would be more appropriate. This is also the case with orthogonal polynomials. You may already know from studying Fourier Transforms (which we will cover in the next chapter), that it can take a very large number of terms to converge for certain shapes, one of these being a step function. The Fourier series for a step (or pulse) converges like 1/n, which is about as slow as it gets.

We will next consider some pulse-like data. It's not a true pulse, because it's not an electronic or mechanical switch. This data comes from a fluorometer, a device for measuring the concentration of fluorescent dye in water, in this chase Rhodamine. The dye is circulated through the fluorometer by a little pump. The output is a 0-5 volt signal, sampled every 10 seconds. It isn't a true pulse because the dye spreads ahead of the flow as well as behind plus the device has a finite response time.

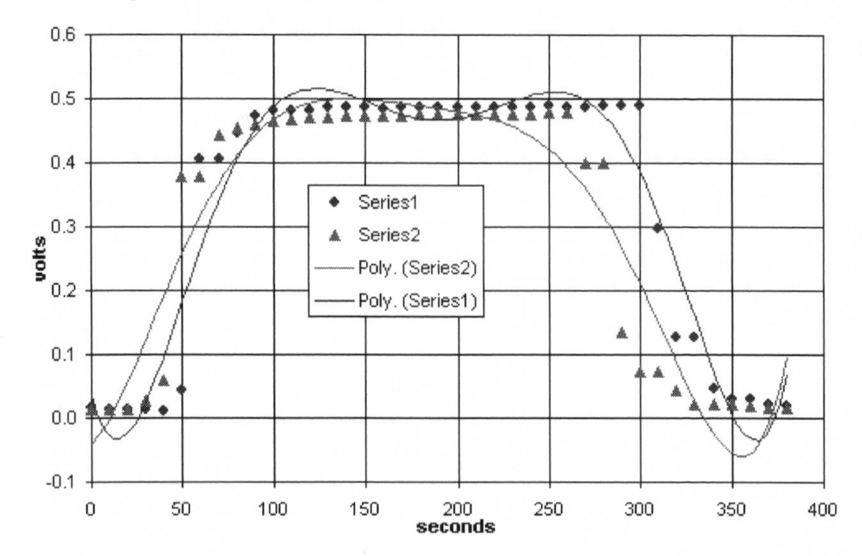

The red data points are measured approximately an hour after the blue ones. The slight reduction in output (volts) is due to degradation of the dye over this time. The thin blue and red curves are the highest order polynomial (6th) that Excel® will fit through these points. Higher order conventional polynomial approximations are possible, but vary wildly and are entirely inappropriate. The same is not true for approximations based on Legendre polynomials. The data and curve fits can be found in pulse.xls, located in the same folder as aperiodic.xls. The 40-term (39th order) Legendre approximation is shown in this next figure:

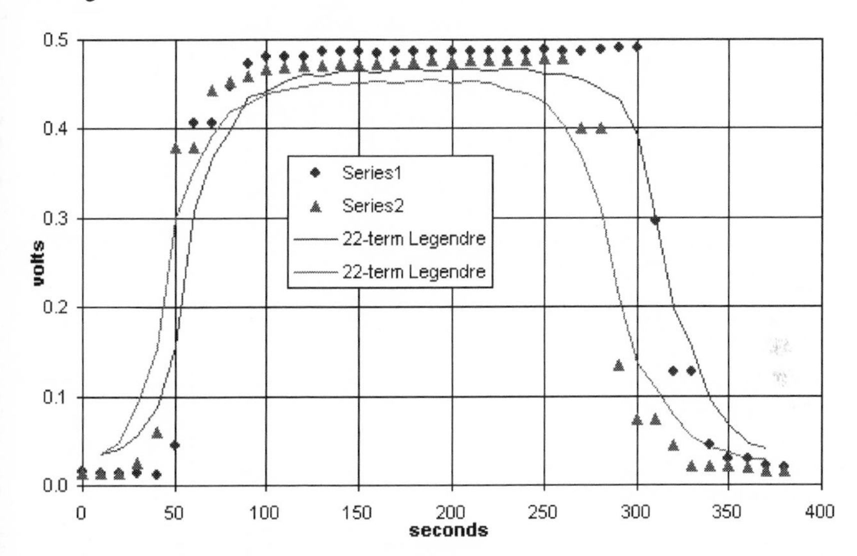

Even with 40 terms, this doesn't adequately represent the data. Perhaps the best option is a 3-point running average, which Excel® will do.

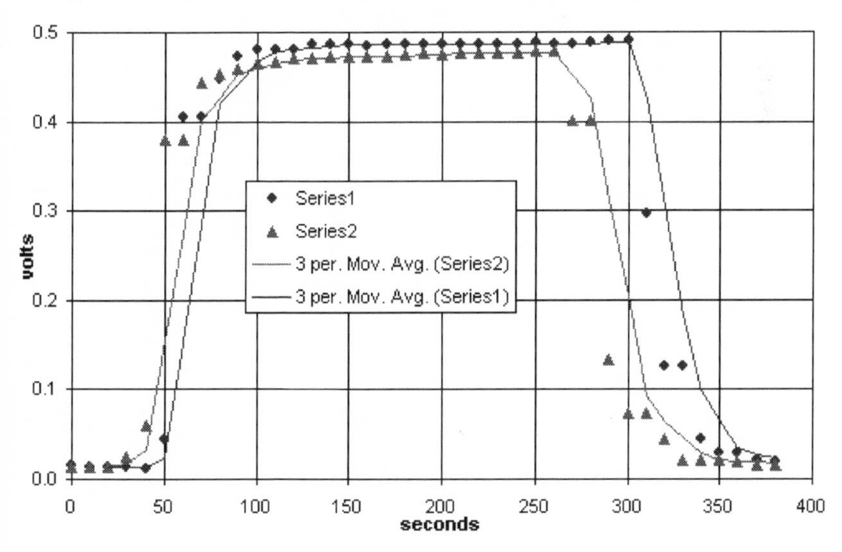

At least this passes through most of the points, although it's not convenient for analysis, plus it's shifted to the right by 15 seconds. This shift is an artifact of Excel®, which performs a running average on the y values, but not the t values, instead associating the average with the last one. The bottom line? Orthogonal functions are not always the best choice, but neither is some ridiculous 50th order polynomial.

Chapter 2. Magic of the Fast Fourier Transform

Surprisingly, many articles about the Fast Fourier Transform (FFT) on the Web skip right over the most astonishing part: where the transcendental values come from in the first place! Perhaps fast floating point processors have been available for so long that these authors have forgotten how many steps it take to calculate a sine or cosine, but I haven't. Sines and cosines form an orthogonal set over the interval $-\pi$ to $+\pi$ (or 0 to 2π); therefore, we can take advantage of this property as we did for the discrete polynomials in Chapter 1.

In order to determine the coefficients for our approximation, we will need to calculate a series of sums. This means we must calculate a whole bunch of sines and cosines. While such may now be possible in microseconds, it hasn't always been this way, which is why FFTs are so useful. Consider the following trigonometric identities:

$$\begin{aligned} \sin(a+b) &= \sin(a)\cos(b)+\cos(a)\sin(b) \\ \cos(a+b) &= \cos(a)\cos(b)-\sin(a)\sin(b) \end{aligned} \tag{2.1}$$

The sums we will be calculating include sin(x), sin(2x), sin(3x), etc. and cos(x), cos(2x), cos(3x), etc. Sin(2x) is equal to sin(x+x) and sin(3x) is equal to sin(2x+x), etc. We need only calculate sin(x) and cos(x), because all the higher order terms come from repeated applications of Equation 2.1. This is the *magic of the FFT!* The code to implement this is almost trivial. The following snippets illustrate the Slow and Fast Fourier Transform. Note that the transcendental functions (i.e., sin and cos) are inside the inner loop in the former and outside in the latter.

```
void SlowFourierTransform(double*f,int
   n,double*c,double*s,int m)
  {
  int i,j;
  double ci,cj,cs,si,sj,ss;
  for(i=0;i<m;i++)
    {
    for(cj=cs=sj=ss=j=0;j<n;j++)
      {
      ci=cos(2*i*(j+1)*M_PI/n);
      si=sin(2*i*(j+1)*M_PI/n);
      cj+=f[j]*ci;
      sj+=f[j]*si;
      cs+=ci*ci;
      ss+=si*si;
      }
    c[i]=cj/cs;
    if(i)
      s[i]=sj/ss;
    }
  }
```

```
void FastFourierTransform(double*f,int
   n,double*c,double*s,int m)
  {
  int i,j;
  double cc,ci,cj,ck,si,sj,sk,ss;
  for(i=0;i<m;i++)
    {
    ci=cj=cos(2*i*M_PI/n);
    si=sj=sin(2*i*M_PI/n);
    for(c[i]=s[i]=cc=ss=j=0;j<n;j++)
      {
      c[i]+=cj*f[j];
      s[i]+=sj*f[j];
      cc+=cj*cj;
      ss+=sj*sj;
      ck=ci*cj-si*sj;
      sk=si*cj+ci*sj;
      cj=ck;
      sj=sk;
      }
    c[i]/=cc;
    if(i)
      s[i]/=ss;
    }
  }
```

If *f(x)* is a periodic function over the interval -π to +π, the Fourier series is defined by:

$$F(x) = \frac{a_0}{2} + \sum_{n=1}^{\infty} a_n \cos(nx) + b_n \sin(nx) \tag{2.2}$$

where the coefficients a_n and b_n are defined by:

$$\begin{aligned} a_n &= \frac{1}{\pi} \int_{-\pi}^{\pi} f(x)\cos(nx)dx \\ b_n &= \frac{1}{\pi} \int_{-\pi}^{\pi} f(x)\sin(nx)dx \end{aligned} \tag{2.3}$$

We will consider three simple examples (step, saw tooth, and ramp) for which the analytical solutions are readily available to illustrate this process and validate the calculations. These are illustrated in the following figure:

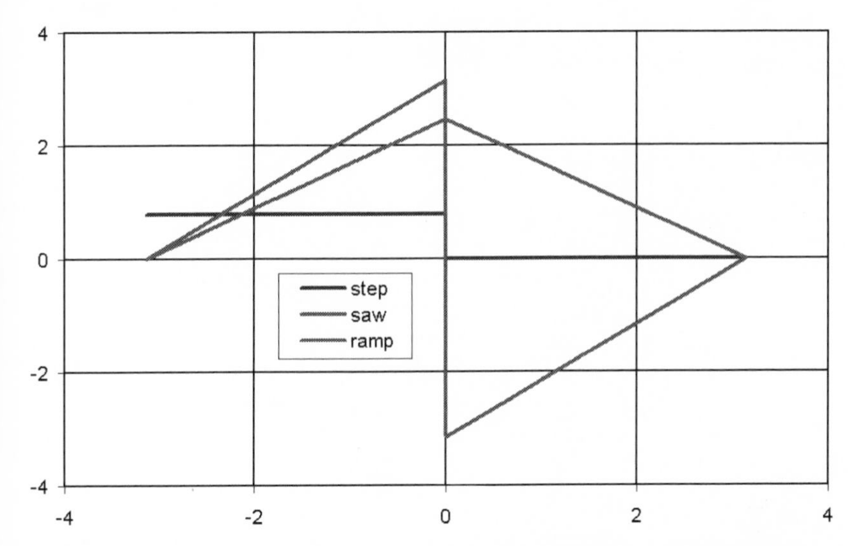

The step series is given by: $a_n=0$, $b_n=1/n$, $n=1,3,5,\ldots$ The preceding C function yields: 0, 1, 0, 0.33332, 0.00001, 0.19998, 0.00001, 0.14284, 0.00002, 0.11108. The saw tooth is given by $a_n=0$, $b_n=((-1)^{n+1})/n$, $n=1, 2, 3, \ldots$ The preceding C function yields: 0, 0.99902, -0.49999, 0.33299, -0.24998, 0.19978, -0.16663, 0.14268, -0.12496, 0.11095. The ramp is given by: $a_0=b_n=0$, $a_n=1/n^2$, $n=1, 3, 5, \ldots$ The preceding C function yields: 0.00181, 1.00121, 0.0012, 0.11232, 0.0012, 0.04121, 0.0012, 0.02161, 0.0012, 0.01355. Output of the code follows, including a comparison of results and processor time:

```
                Square Wave
     <-----slow----->  <-----fast----->
i     c[i]      s[i]      c[i]      s[i]
0   0.00016  0.00000  0.00016  0.00000
1  -0.00094  1.00000 -0.00094  1.00000
2   0.00031  0.00000  0.00031  0.00000
3  -0.00094  0.33333 -0.00094  0.33333
4   0.00031  0.00000  0.00031  0.00000
5  -0.00094  0.20000 -0.00094  0.20000
6   0.00031  0.00000  0.00031  0.00000
7  -0.00094  0.14285 -0.00094  0.14285
8   0.00031  0.00000  0.00031  0.00000
9  -0.00094  0.11111 -0.00094  0.11111
    0.05480  seconds  0.00448  seconds
                 Saw Tooth
     <-----slow----->  <-----fast----->
i     c[i]      s[i]      c[i]      s[i]
0   0.00031  0.00000  0.00031  0.00000
1  -0.00126  0.99960 -0.00126  0.99960
2   0.00126 -0.50000  0.00126 -0.50000
```

```
3 -0.00126  0.33320 -0.00126  0.33320
4  0.00126 -0.25000  0.00126 -0.25000
5 -0.00126  0.19992 -0.00126  0.19992
6  0.00126 -0.16666  0.00126 -0.16666
7 -0.00126  0.14279 -0.00126  0.14279
8  0.00126 -0.12499  0.00126 -0.12499
9 -0.00126  0.11106 -0.00126  0.11106
   0.05460  seconds  0.00446  seconds
                 Ramp
   <-----slow----->  <-----fast----->
i    c[i]     s[i]     c[i]     s[i]
0  0.00049  0.00000  0.00049  0.00000
1  1.00000  0.00063  1.00000  0.00063
2 -0.00000 -0.00000 -0.00000 -0.00000
3  0.11111  0.00021  0.11111  0.00021
4 -0.00000 -0.00000 -0.00000 -0.00000
5  0.04000  0.00013  0.04000  0.00013
6 -0.00000 -0.00000 -0.00000 -0.00000
7  0.02041  0.00009  0.02041  0.00009
8 -0.00000 -0.00000 -0.00000 -0.00000
9  0.01235  0.00007  0.01235  0.00007
   0.05328  seconds  0.00446  seconds
```

The Radon Transform and Medical Imaging

Perhaps the most fascinating application of the FFT is medical imaging, including CAT[5] or MRI scans. A beam is passed through the patient, producing a time-varying analog signal. The signal is reduced to frequency components on arrival by performing an FFT. This also digitizes the signal—all digitally with modern computers, but early implementations used a mixture of analog and digital strategies to accomplish essentially the same result. The sensor is rotated slightly and another beam is passed through the patient. This sensor positioning naturally produces data in polar coordinates, rather than rectangular (i.e., Cartesian). This is called the Radon Transform after Austrian mathematician Johann Karl August Radon (1887-1956).

The conventional polar transform is based on radius, r, and angle theta, θ, and spans $r=0$ to ∞ and $\theta=0$ to 2π. The Radon Transform uses $r=-1$ to $+1$ and $\theta=0$ to π. Otherwise, it's the same. An image often used as a test case is called the Shepp-Logan Phantom and may be found on page iv, immediately following the Table of Contents. This standard image was created by Larry Shepp and

[5] The acronym CAT stands for Computer Aided Tomography and often refers to a rotating X-ray scanner. Magnetic Resonance Imaging is actually a type of CAT scan, although it uses a powerful magnetic field instead of X-rays to sample the patient. A Positron Emission Tomography (PET) scanner is yet another type of Computer Aided Tomography, which uses particles instead of X-rays or magnetic fields.

Benjamin Logan to test their image reconstruction algorithms.[6] A graphical representation of the FFT is shown just below the phantom.

There are many articles and much software on the Web related to this important subject. Sadly, most of the code is incompatible with Windows® or requires some arcane platform such as Matlab®. Peter Toft is an excellent source of information and code. I highly recommend reading his dissertation (which is available at his Web site) if you are interested in this subject.[7] He has provided a suite of programs that are easily adapted to the Windows® O/S. You will find the adaptations in the examples\Recon2D folder, along with batch files to create the executables plus a little program to convert the floating-point output files to bitmap images. The Recon2D.tar.gz archive includes a program to create various phantoms. The standard phantom results are shown below:

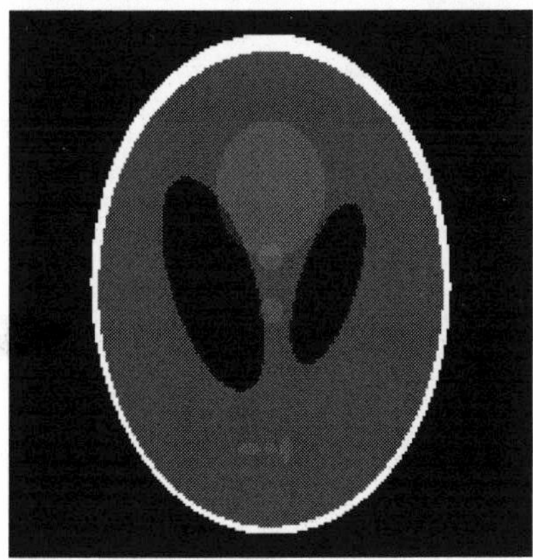

phantom

---

[6] Shepp, L. A., and Logan, B. F., "The Fourier Reconstruction of a Head Section" IEEE Transactions on Nuclear Science, NS-21 (3), pp. 21–43, 1974.

[7] Toft, P., "The Radon Transform - Theory and Implementation," Ph.D. Thesis, Department of Mathematical Modeling, Technical University of Denmark, 1996. visit his Web page at: http://petertoft.dk/PhD/

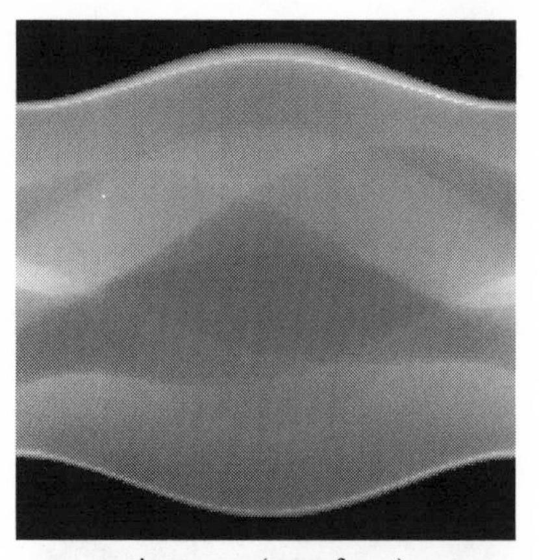

sinogram (transform)

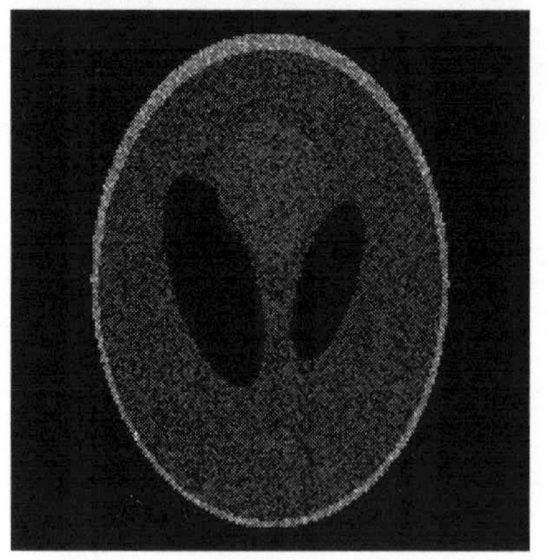

inversion

<u>Touch Tones</u>

The beeps used by touch tone phones are actually two separate pure tones combined together to make a composite sound that is easily distinguished from noise. The following table shows how four tones are combined in pairs to create the sound for each key:

Touch Tone Pairs

		high			
low	Hz	1209	1336	1477	1633
	697	1	2	3	A
	770	4	5	6	B
	852	7	8	9	C
	941	*	0	#	D

The 1 tone is composed of 697 Hz plus 1209 Hz. Graphically this looks like the following:

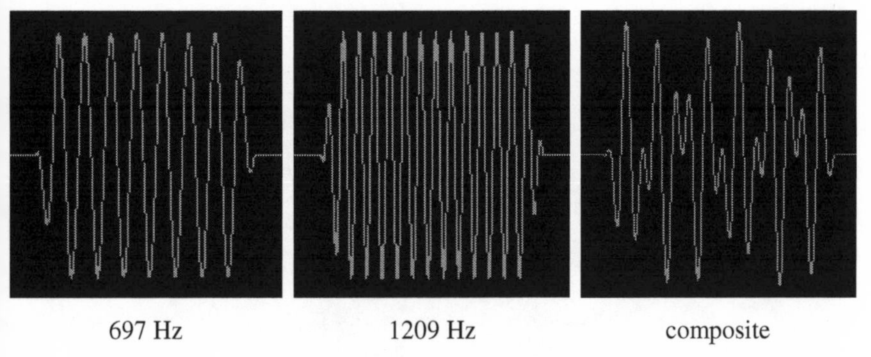

697 Hz 1209 Hz composite

Since we know what the two frequencies are and we want them to have the same volume (i.e., magnitude), we know what the Fourier Transform (FT) is from the outset. We can construct the composite tone directly from the FT. The code to accomplish this is Inverse Fourier Transform (IFT) quite compact:

```
short F1[]={697,770,852,941};
short F2[]={1209,1336,1477};
i=index("123456789*0#",key);
f1=F1[i/4];
f2=F2[i%3];
da=2*f1*M_PI/samp;
yk=2*cos(da);
y1=sin(-2*da);
y2=sin(-da);
da=2*f2*M_PI/samp;
xk=2*cos(da);
x1=sin(-2*da);
x2=sin(-da);
for(l=samp/10;l>0;l--)
  {
  y3=yk*y2-y1;
  y1=y2;
  y2=y3;
  x3=xk*x2-x1;
  x1=x2;
  x2=x3;
  buffer[w++]=(BYTE)((unsigned int)
    (silence+(x3+y3)*loudness));
  }
```

This is the simplest possible sound application of the FFT. You will find a complete but simple Windows® program (playit.c) in Appendix D that generates and plays these and other tones, as well as creates output .WAV files. You will also find a program (wav2csv.c) in folder examples\WAVs that will read any .WAV file and write the data to a .CSV file, which will import directly into Excel®. The touch tone WAVs are also in this folder. Several programs are available on the Web to process WAVs. My personal favorite is Gold Wave by Chris Craig. This is what a typical WAV looks like:

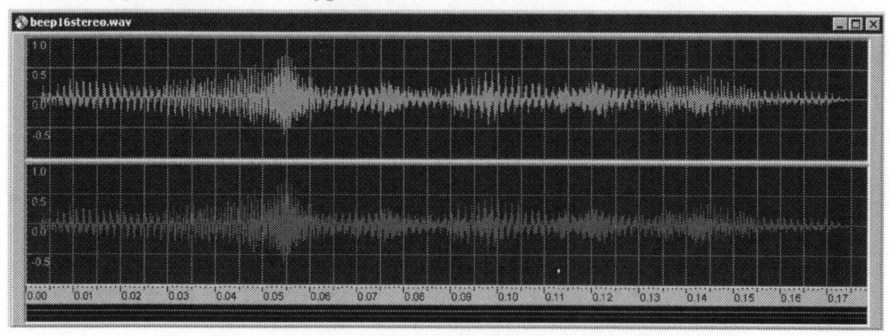

There is also a little program (wav2fft.c) in the examples\WAVs folder that will read a WAV and output the FFT. The above signal produces the following:

```
wav2fft beep16mono
convert .wav to FFT
reading wav file: beep16mono.wav
tag=RIFF
  data block size=15860
tag=WAVE
tag=fmt
  header size=16
  code=1
  channels=1 (mono)
  sampling rate=44100
  sampling rate*channels*bytes/sample=88200
  channels*bytes/sample=2
  bits/sample=16
tag=data
  bytes=15744
transforming sound to floating-point signal
applying FFT
  i      c[i]     s[i]
  123  0.00767  0.02928
  124 -0.04085 -0.02973
  126  0.00818 -0.02602
  247  0.02982  0.00727
  248 -0.02583  0.01227
  249 -0.03390 -0.03033
  251  0.03619  0.03949
  994 -0.02543  0.00898
```

Chapter 3. Orthogonal Polynomials

The discrete orthogonal polynomials introduced in Chapter 1 are interesting, but you probably won't see these anywhere else because nobody uses them. The seven most commonly used orthogonal polynomials, the intervals, and corresponding weighting functions are listed in the following table:

Table 3.1 Most Common Orthogonal Polynomials

name	interval	w(x)
Fourier (sin,cos)	$-\pi$ to π	1
Legendre	-1 to 1	1
Jacobi	-1 to 1	$(1-x)^a(1+x)^b$
Gebenbauer (ultraspherical)	-1 to 1	$(1-x^2)^{a-½}$
Chebyshev (1st kind)	-1 to 1	$(1-x)^{-½}$
Chebyshev (2nd kind)	-1 to 1	$(1-x)^{½}$
Laguerre	0 to ∞	$e^{-x}x^a$
Hermite	$-\infty$ to ∞	e^{-x^2}

Sin and cos aren't polynomials, but are included in the table to illustrate two things: the interval and the weighting function. The orthogonality condition is:

$$\int_a^b P_i(x)P_j(x)w(x)dx = \begin{cases} \neq 0 \, if \; i = j \\ = 0 \, if \; i \neq j \end{cases} \tag{3.1}$$

The interval is different for the Fourier series and the Legendre polynomials, but these are the only ones with no weighting function (i.e., $w(x)=1$). All the others have a weighting function that varies over the interval. The Laguerre and Hermite are the only ones with infinite intervals, the former semi-infinite and the latter fully infinite. The rest operate over a finite domain. Which ones to use is determined by the domain and the weighting function. Select the ones that correspond to your problem. More details about these and other functions may be found in Chapter 22 (especially sections 22.18-22.20) of the seminal reference for applied mathematicians, Abramowitz and Stegun.[8]

Aperiodic Transient Data

We will first consider a case of transient data that are known to be aperiodic (i.e., not periodic—don't repeat over and over again at any regular time interval). There are five temperature sensors bound together with a cable tie. These should all be reading the same temperature, but they don't. These are high-grade platinum 4-wire resistance thermal devices (RTDs) purchased from Omega®, the premier supplier of such instruments, for about $100/ea. The sensor response

---

[8] Abramowitz, M. and I. A. Stegun, *Handbook of Mathematical Functions* first published by the National Bureau of Standards as Technical Monograph No. 55. This invaluable reference may be obtained free online as a PDF from several different web sites.

is actually ohms, as nothing outputs temperature, as such. Resistance in ohms is related to temperature in degrees C via the Callendar-Van Dusen Equation:

$$R = D\left[1 + AT + BT^2 + (T - 100)CT^3\right] \tag{3.2}$$

The five reported temperatures are shown in this next figure:

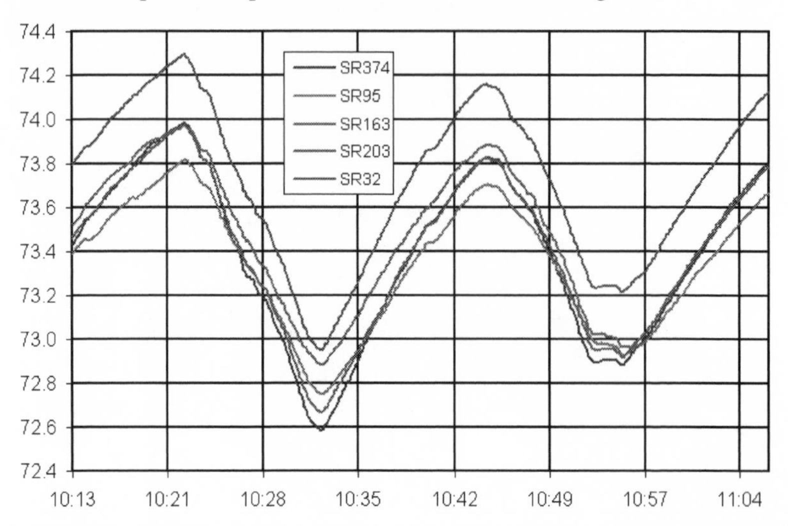

Clearly, there is a common phenomenon driving the response of all five sensors. There appears to be an offset, which would be a calibration error. We will use orthogonal polynomials to extract the common information and eliminate the noise. As there is no clear weighting function and this is not a periodic signal, the appropriate choice is Legendre. The first 10 are:

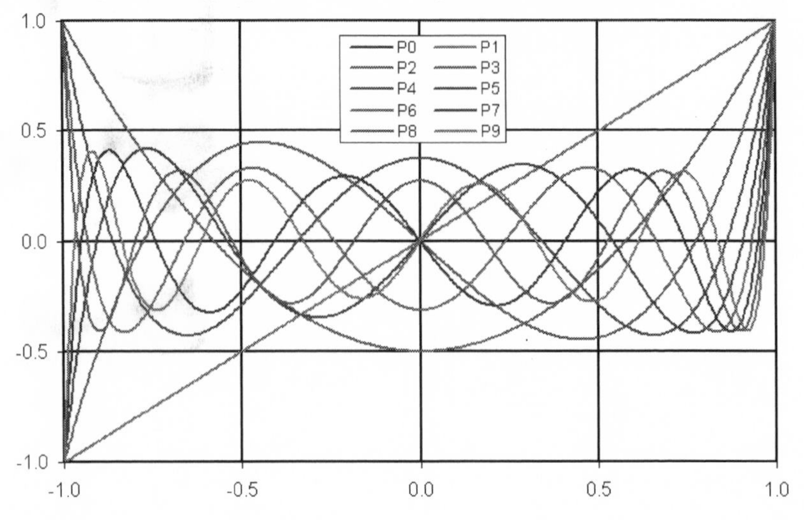

If we simply sum the products of the data and polynomials, as is often done with Fourier Transforms and was done with the discreetly orthogonal polynomials in Chapter 1, the result is:

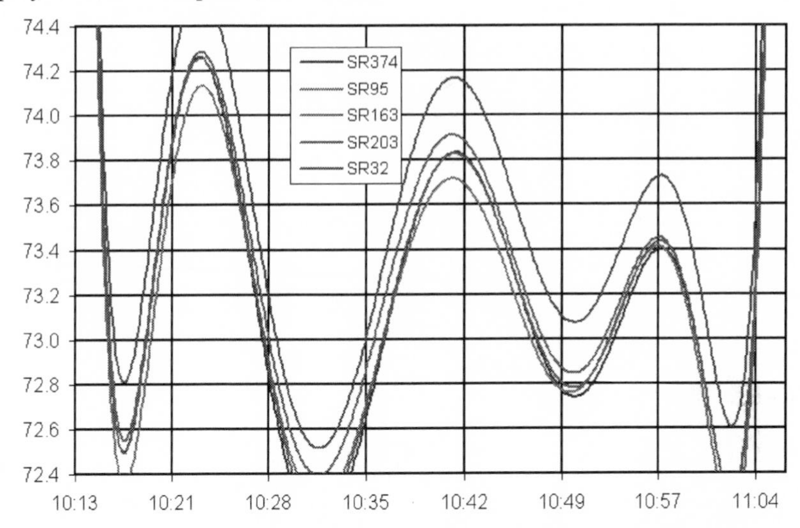

These curves don't fit the data well at all. What's the problem here?

Pivoting vs. Projecting

The length of the shadow a vector casts on the X-, Y-, or Z-axis is the projection of that vector onto each of the mutually perpendicular axes, as illustrated in the following figure. This is in effect what we do by summing data and orthogonal functions.

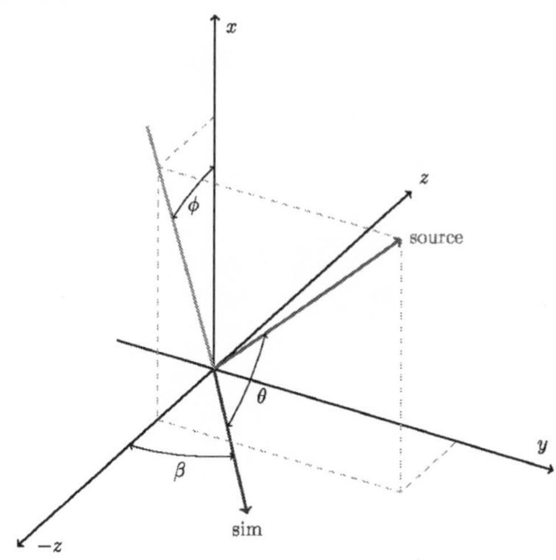

Ideally, this would work, but reality is more complicated. There's noise in most data and we can't ignore it. We also can't sample (or integrate) infinitesimally. Data always have limited resolution. We compensate for this by *pivoting*. After each sum, we remove the contribution of that polynomial (or vector component) from the data. This simple step has remarkable results because the subsequent functions are orthogonal to the components we remove. Projected and pivoted are shown in this next figure for the first probe.

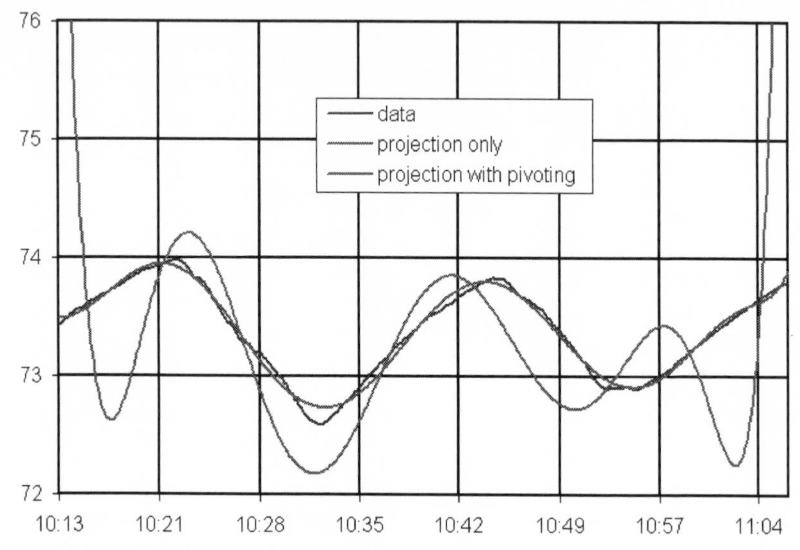

The projected red curve doesn't fit well at all; whereas, the pivoted curve fits quite well. This is easy to implement in code. The subtle differences are shown below in bold.

```
void ProjectData()
  {
  int i,k,l;
  double p,s,T,x;
  for(l=0;l<5;l++)
    {
    for(i=0;i<m;i++)
      {
      for(a[l][i]=s=k=0;k<data.n;k++)
        {
        x=-1.+2.*(data.t[k]-data.t[0])/
         (data.t[data.n-1]-data.t[0]);
        T=data.T[l][k];
        p=Legendre(i,x);
        a[l][i]+=p*T;
        s+=p*p;
        }
```

```
      a[l][i]/=s;
      }
    }
  }
void PivotData()
  {
  int i,j,k,l;
  double p,s,T,x;
  for(l=0;l<5;l++)
    {
    for(i=0;i<m;i++)
      {
      for(b[l][i]=s=k=0;k<data.n;k++)
        {
        x=-1.+2.*(data.t[k]-data.t[0])/
          (data.t[data.n-1]-data.t[0]);
        T=data.T[l][k];
        for(j=0;j<i;j++)
          T-=b[l][j]*Legendre(j,x);
        p=Legendre(i,x);
        b[l][j]+=p*T;
        s+=p*p;
        }
      b[l][i]/=s;
      }
    }
  }
```

The results with pivoting for all 5 probes is:

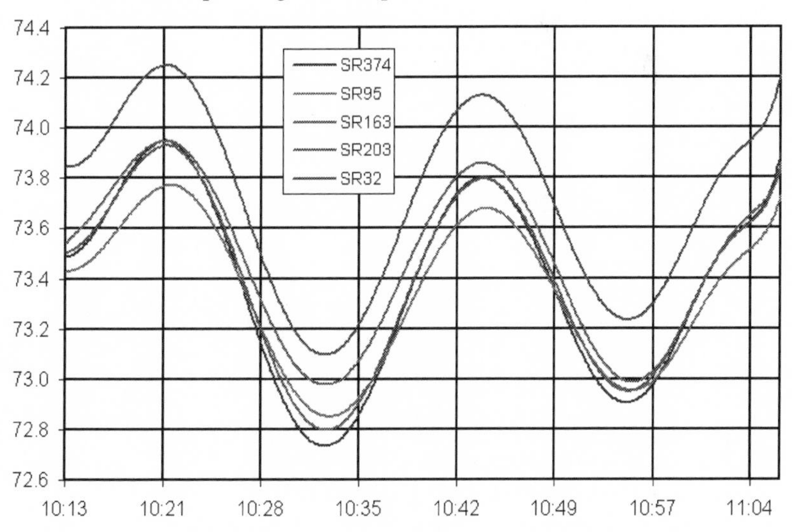

The correlated response is now clear, as is the discrepancy in calibration.

```
aperiodic
reading data file: aperiodic.csv
  437 lines read
  436 points found
projecting data
  a[0]={73.3815,-0.0918717,1.11625,0.0377154,1.58683,
       0.929226,1.70563,-0.466497,3.00876,0.466632};
  a[1]={73.3412,-0.0795873,1.02169,0.0578899,1.54963,
       0.76912,1.85602,-0.35623,2.90657,0.423041};
  a[2]={73.4716,-0.125402,1.01525,0.0713992,1.62531,
       0.823994,1.77014,-0.38697,2.91555,0.42581};
  a[3]={73.4013,-0.0921603,1.09157,0.0377137,1.57166,
       0.883002,1.75174,-0.438666,2.97386,0.455708};
  a[4]={73.7138,-0.0960703,1.0991,0.0435049,1.61411,
      0.901204,1.73693,-0.456192,3.00717,0.459874};
pivoting data
  b[0]={73.3815,-0.126617,0.309062,-0.0419117,0.132731,
       0.803636,-0.399201,-0.666291,0.267469,0.236591};
  b[1]={73.3412,-0.114313,0.214938,-0.021777,0.0981434,
       0.643044,-0.243703,-0.551725,0.165742,0.193902};
  b[2]={73.4716,-0.16019,0.207102,-0.00769773,0.171454,
       0.698337,-0.335228,-0.58351,0.170883,0.196677};
  b[3]={73.4013,-0.126915,0.284166,-0.041903,0.117667,
       0.757455,-0.352406,-0.636981,0.231598,0.226411};
  b[4]={73.7138,-0.130973,0.288262,-0.0363998,0.153868,
       0.775001,-0.377316,-0.655936,0.25267,0.229484};
filing comparison: comparison.csv
filing calibrated results: calibrated.csv
```

The corrected results without noise are shown in this next figure. All of the associated files can be found in folder examples\aperiodic.

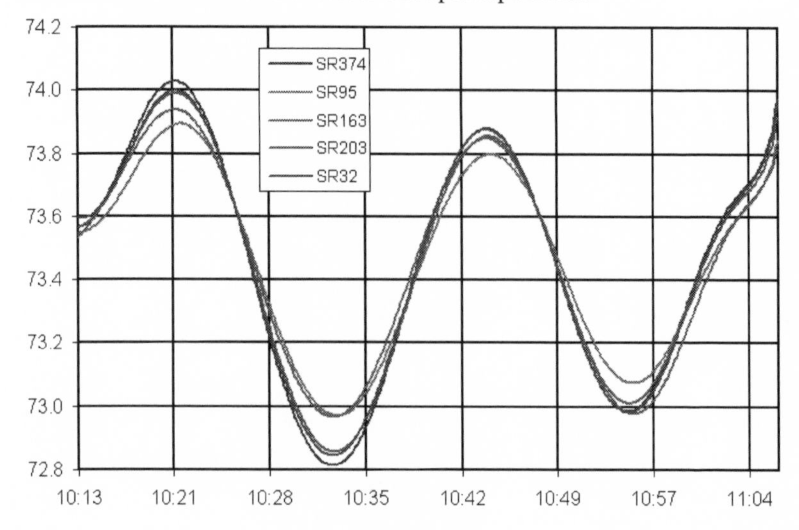

Chebyshev Polynomials and Simple Weighed Data

There was no reason to weight the preceding data, but this is not always the case. In addition to dealing with noise, sometimes we care more about the data on the ends than in the middle (or vise versa). We will next approximate some data using Chebyshev polynomials of either the 1st or 2nd kind. The associated weighting functions are listed in the preceding table (i.e., *1/sqrt(1-x^2)* and *sqrt(1-x^2)*, respectively). The former will emphasize data on ends; whereas, the latter will emphasize data in the middle. The spreadsheets and code can be found in the folder examples\Chebyshev. The 1st kind are shown in the following figure:

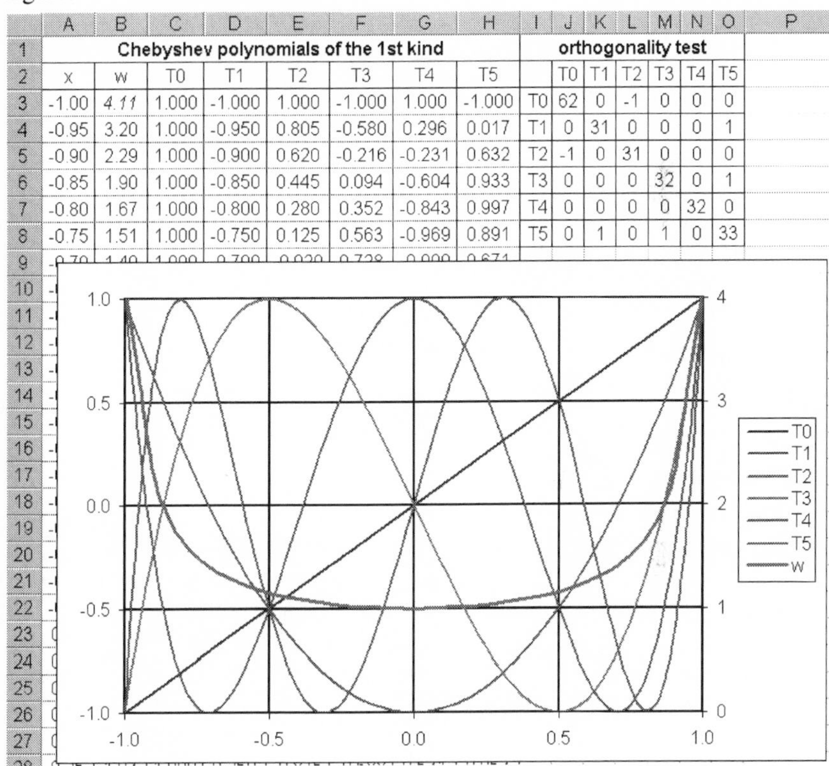

Chebyshev polynomials of the 1st kind								orthogonality test						
x	w	T0	T1	T2	T3	T4	T5		T0	T1	T2	T3	T4	T5
-1.00	*4.11*	1.000	-1.000	1.000	-1.000	1.000	-1.000	T0	62	0	-1	0	0	0
-0.95	3.20	1.000	-0.950	0.805	-0.580	0.296	0.017	T1	0	31	0	0	0	1
-0.90	2.29	1.000	-0.900	0.620	-0.216	-0.231	0.632	T2	-1	0	31	0	0	0
-0.85	1.90	1.000	-0.850	0.445	0.094	-0.604	0.933	T3	0	0	0	32	0	1
-0.80	1.67	1.000	-0.800	0.280	0.352	-0.843	0.997	T4	0	0	0	0	32	0
-0.75	1.51	1.000	-0.750	0.125	0.563	-0.969	0.891	T5	0	1	0	1	0	33

The weighting function is the thick magenta curve.

The 2nd kind are shown in this next figure:

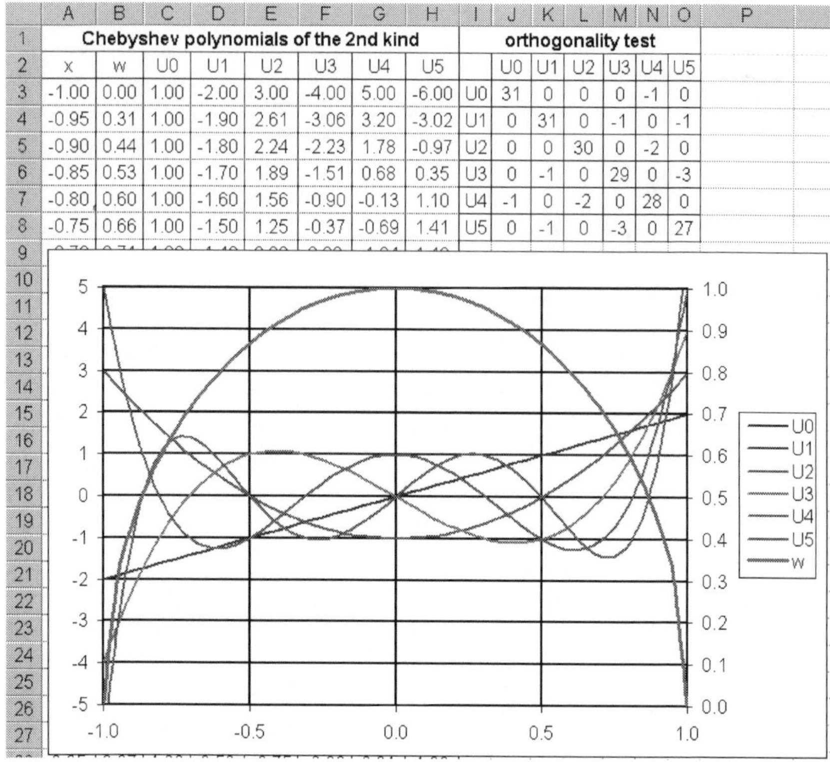

	A	B	C	D	E	F	G	H	I	J	K	L	M	N	O
1	Chebyshev polynomials of the 2nd kind								orthogonality test						
2	x	w	U0	U1	U2	U3	U4	U5		U0	U1	U2	U3	U4	U5
3	-1.00	0.00	1.00	-2.00	3.00	-4.00	5.00	-6.00	U0	31	0	0	0	-1	0
4	-0.95	0.31	1.00	-1.90	2.61	-3.06	3.20	-3.02	U1	0	31	0	-1	0	-1
5	-0.90	0.44	1.00	-1.80	2.24	-2.23	1.78	-0.97	U2	0	0	30	0	-2	0
6	-0.85	0.53	1.00	-1.70	1.89	-1.51	0.68	0.35	U3	0	-1	0	29	0	-3
7	-0.80	0.60	1.00	-1.60	1.56	-0.90	-0.13	1.10	U4	-1	0	-2	0	28	0
8	-0.75	0.66	1.00	-1.50	1.25	-0.37	-0.69	1.41	U5	0	-1	0	-3	0	27

The weighting function is the thick magenta curve.

The interval for both types is -1 to +1, so we will apply what is called a *odd reflection*. An even reflection would be $f(-x)=f(x)$; whereas, an odd reflection would be $f(-x)= -f(x)$. This means we will first mirror the data right-to-left, and then flip it top-to-bottom. We will also insert a point at (0,0). While it's not practical to measure data at $x=0$ in this case, we know the result must be $f(0)=0$, because this would correspond to nothing happening (i.e., equilibrium).

The original plus reflected data are shown in the following figure:

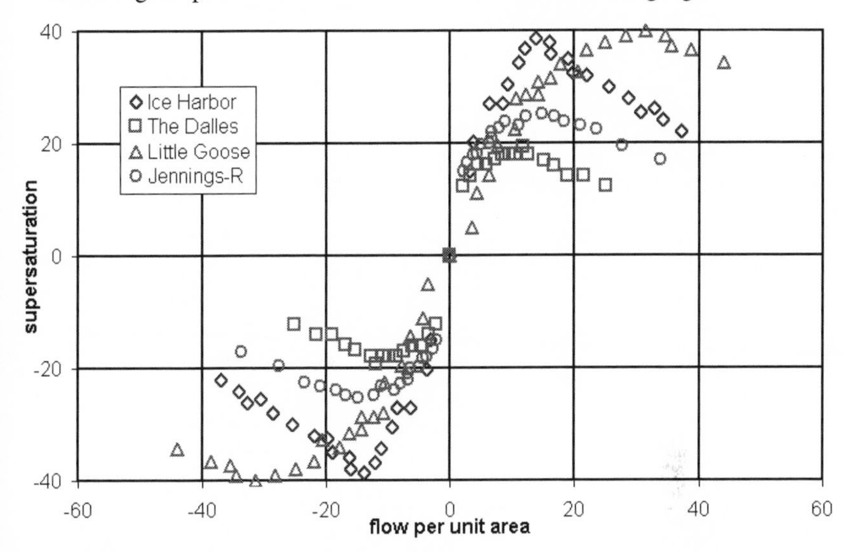

We perform the same process as with the FFT (i.e., pivoting), only we use either the 1st or 2nd Chebyshev polynomial along with the associated weight. We only use the odd orders because we performed the odd reflection and the even orders should all degenerate. The code is quite simple:

```
for(i=0;i<m;i++)
  {
  for(c[i]=s=k=0;k<n;k++)
    {
    if(k<data->n) /* odd reflection */
      {
      xk=-data->x[data->n-1-k];
      yk=-data->y[data->n-1-k];
      }
    else if(k>data->n) /* original data */
      {
      xk=data->x[k-1-data->n];
      yk=data->y[k-1-data->n];
      }
    else /* insert point (0,0) */
      xk=yk=0.;
    t=xk/r;
    for(j=0;j<i;j++)
      yk-=c[j]*Cheby1(2*j+1,t);
    w=1./sqrt(1.-t*t);
    p=Cheby1(2*i+1,t);
    s+=w*p*p;
    c[i]+=w*p*yk;
```

```
    }
  c[i]/=s;
  }
```

Five terms are adequate. We know this because they just keep getting smaller, which is why we used orthogonal polynomials in the first place—so that we'd know when to stop. Results using the 1st kind are shown below:

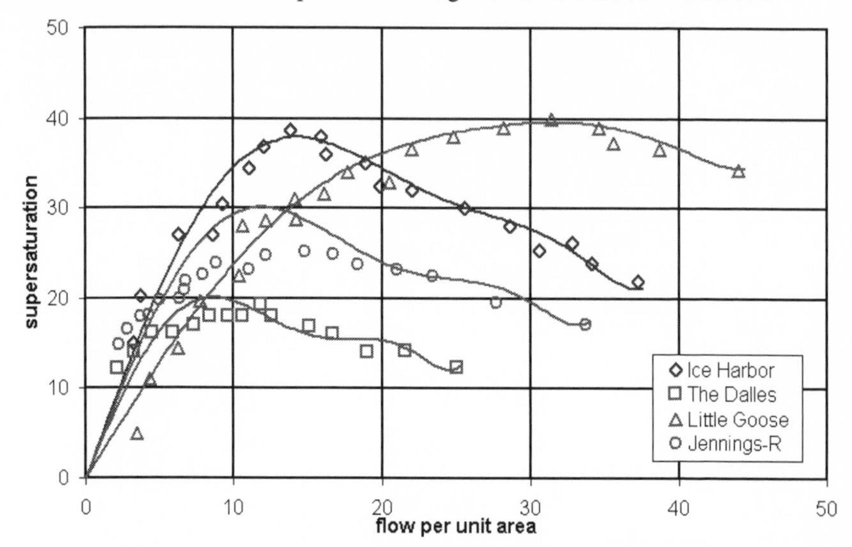

All is well… we've eliminated the noise without biasing the data. Results for the 2nd kind are shown in this next figure:

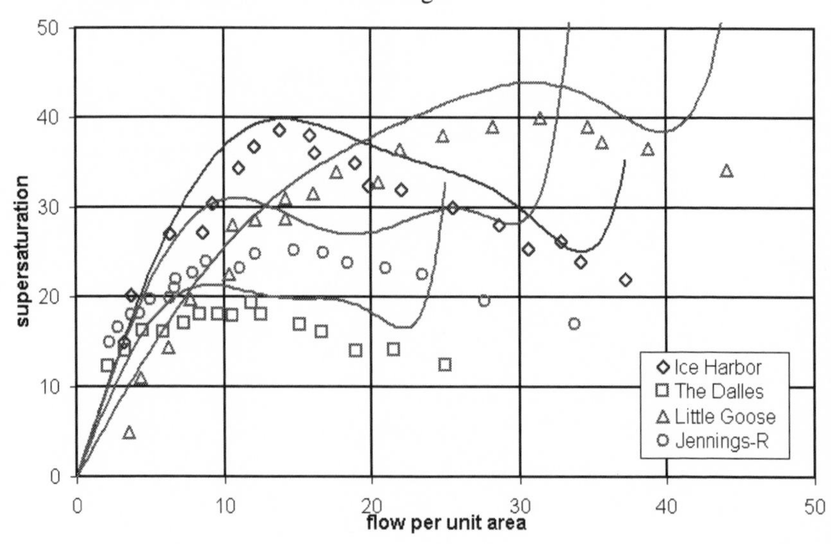

Here we see the impact of the two different weighting functions. The 1st kind emphasized the data on the right (away from zero); whereas, the 2nd kind emphasized the data on the left (closer to zero). The penalty paid for trying to tighten up the regression near zero is unacceptable behavior for large values of x. This is a chronic problem with curve fitting. A high-order regression may fit the data quite well, but often has *tails* (very high or very low values) that quickly depart from the perceived trend in the data that we would draw in if we were extending the curve by hand. Appropriate weighting is how you eliminate this problem. Orthogonal functions are how you avoid biasing the shape of the data, which is a passive form of cherry picking.

Jacobi Polynomials and Skew Weighed Data

Several of the orthogonal functions listed previously are special cases of the Jacobi polynomials. These have two additional parameters (α and β) that can be adjusted to skew the weight to one side or the other (high or low). You will find these in C and Excel® in the folder examples\Jacobi. This is what they look like for one choice of α and β. Notice the skewed weighting function (heavy magenta curve):

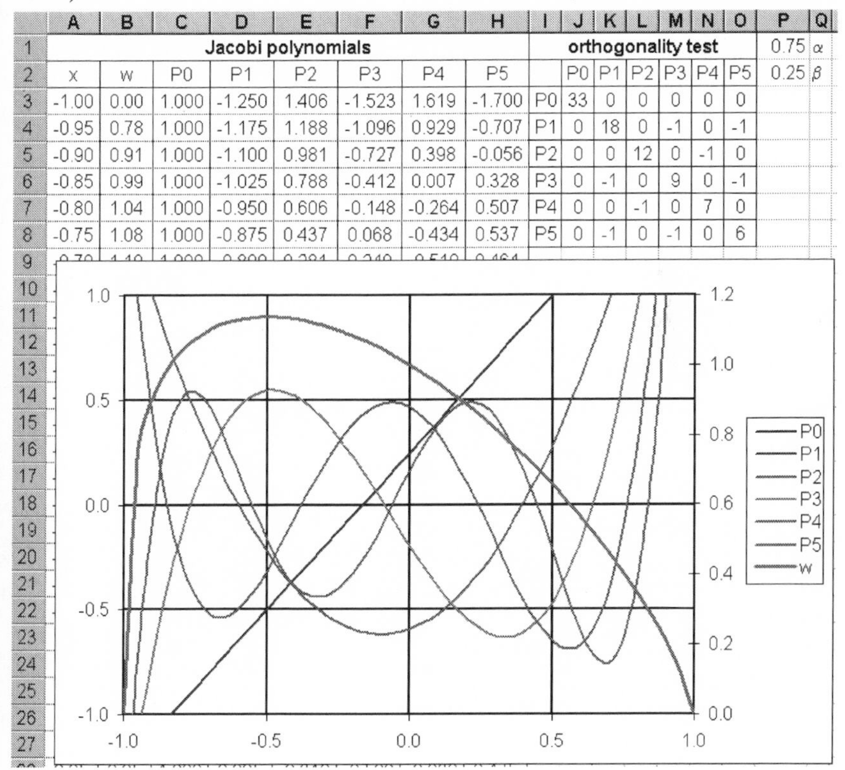

	A	B	C	D	E	F	G	H	I	J	K	L	M	N	O	P	Q
1	Jacobi polynomials								orthogonality test							0.75	α
2	x	w	P0	P1	P2	P3	P4	P5		P0	P1	P2	P3	P4	P5	0.25	β
3	-1.00	0.00	1.000	-1.250	1.406	-1.523	1.619	-1.700	P0	33	0	0	0	0	0		
4	-0.95	0.78	1.000	-1.175	1.188	-1.096	0.929	-0.707	P1	0	18	0	-1	0	-1		
5	-0.90	0.91	1.000	-1.100	0.981	-0.727	0.398	-0.056	P2	0	0	12	0	-1	0		
6	-0.85	0.99	1.000	-1.025	0.788	-0.412	0.007	0.328	P3	0	-1	0	9	0	-1		
7	-0.80	1.04	1.000	-0.950	0.606	-0.148	-0.264	0.507	P4	0	0	-1	0	7	0		
8	-0.75	1.08	1.000	-0.875	0.437	0.068	-0.434	0.537	P5	0	-1	0	-1	0	6		

Larger values of α skew to the left; whereas, larger values of β skew to the right. Smaller values of α and β flatten out the weight (more evenly distributed across the interval); whereas, larger values of α and β produce a more peaked weight (more emphasis on data near the center). You can change the values and the spreadsheet will update automatically.

Chapter 4. Infinite Domains

While most data is confined to some finite extent of time and/or space, some data are best considered as occupying an infinite domain. The first example of this we will consider is contamination. While the extent of such is not literally infinite, we must consider the possibility that the substance in question could spread into a vast region and persist over many lifetimes. We always begin with some measurements, as this is always the first step in assessing the risk. These measurements will necessarily contain some level of noise. While we might fit a generic curve to the data, orthogonal functions enable us to eliminate the noise without excessively biasing the outcome.

For an infinite one-dimensional domain the choice is Hermite polynomials. The associated weighting function is exp(-x2). The Hermite polynomials satisfy several differential equations, two of which are given below:

$$\frac{d^2y}{dx^2} + 2x\frac{dy}{dx} + 2ny = 0 \quad y = \sum CnHn(x) \tag{4.1}$$

$$\frac{d^2y}{dx^2} + \left(2n + 1 - x^2\right)y = 0 \quad y = \sum CnHn(x)e^{\frac{-x^2}{2}} \tag{4.2}$$

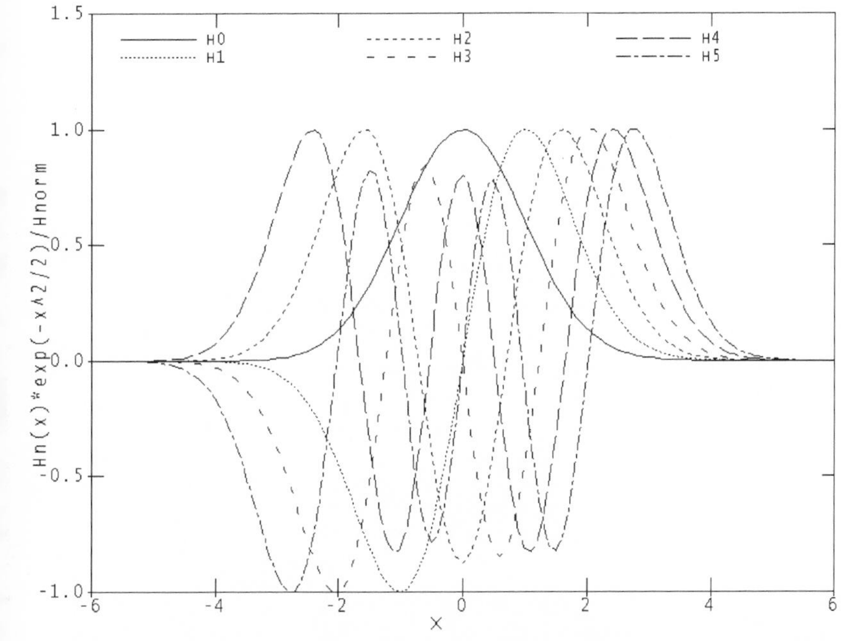

The first 6 orders normalized with the weighting function are shown in the figure above. The Hermite polynomials are orthogonal with respect to the weighting function exp(-x2/2) over the doubly infinite range (viz. -∞to +∞) as indicated by the following integral:

$$\int_{-\infty}^{+\infty} Hn(x)Hm(x)e^{-x^2}\,dx \;=\begin{cases}\sqrt{\pi 2^n n!} & if(n=m)\\ 0 & if(n\neq m)\end{cases} \tag{4.3}$$

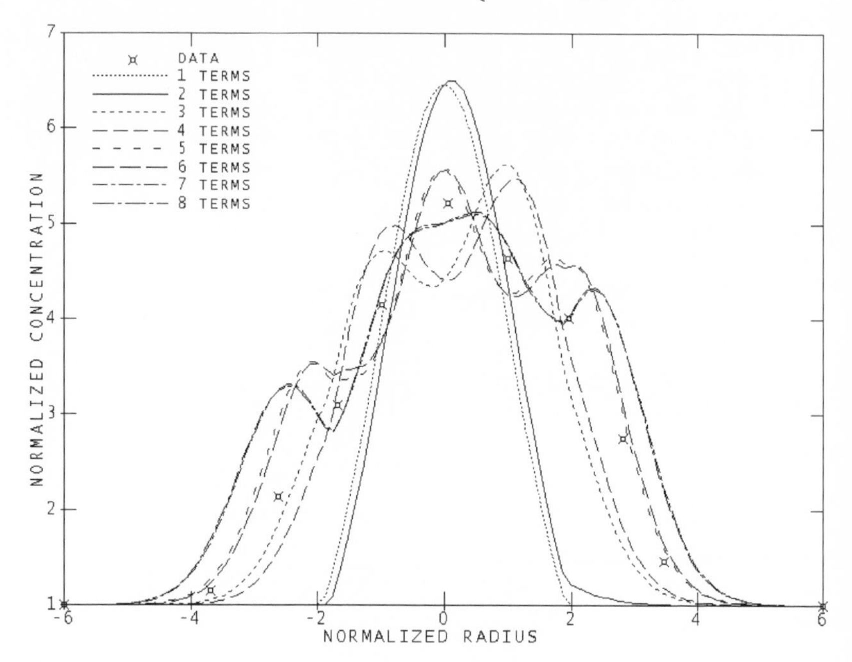

Hermite Approximation

The preceding figure shows the successive orders of the Hermite series approximation for a test data set. This is similar to the successive order of the Fourier expansion for a function in the time domain as well as the Chebyshev and Jacobi polynomials in Chapter 3. The terms in the series are computed from the orthogonality conditions as follows:

$$Cn = \frac{\int_{-\infty}^{+\infty} F(x)Hn(x)e^{-x^2}\,dx}{\int_{-\infty}^{+\infty} H_n^2(x)e^{-x^2}\,dx} \tag{4.4}$$

Test of Integration the Techniques

A test of the integration methods follows by applying the technique to the orthogonality condition (Equation 3). If you generate a sequence of data sets, which are the various products of the Hermite functions, the accuracy of the integration technique can be determined by checking the residual matrix whose elements are defined as follows:

$$Rn = \frac{\int_{-\infty}^{+\infty} Hn(x)Hm(x)e^{-x^2}\,dx}{\int_{-\infty}^{+\infty} H_n^2(x)e^{-x^2}\,dx} \tag{4.5}$$

The polynomials in C code (Hermite.c, fig1.c, fig2.c) and Excel® (Hermite.xls) can be found in the folder examples\Hermite. You will also find three different approaches to numerical integration over an infinite domain (gquad.c) in this folder.[9] The first of these codes (Hermite.c) demonstrates the orthogonality property. Abbreviated output follows:

```
\examples\Hermite>hermite
illustrating the orthogonality of Hermite polynomials using GQ96II
         0         1         2         3         4         5         6
0     1.772     0.000    -0.000     0.000    -0.000     0.000     0.000
1     0.000     3.545     0.000    -0.000     0.000    -0.000     0.000
2    -0.000     0.000    14.180     0.000    -0.000     0.000    -0.000
3     0.000    -0.000     0.000    85.078     0.000    -0.000     0.000
4    -0.000     0.000    -0.000     0.000   680.622     0.000     0.000
5     0.000    -0.000     0.000    -0.000     0.000  6806.223     0.000
6     0.000     0.000    -0.000     0.000     0.000     0.000 81674.673
illustrating the orthogonality of Hermite polynomials using GQ999IIa
         0         1         2         3         4         5         6
0     0.886     1.000    -0.000    -2.000     0.000    12.000    -0.000
1     1.000     1.772     2.000     0.000    -4.000    -0.000    24.000
2    -0.000     2.000     7.090    12.000    -0.000   -40.000    -0.000
3    -2.000     0.000    12.000    42.539    72.000    -0.000  -240.000
4     0.000    -4.000    -0.000    72.000   340.311   720.000    -0.000
5    12.000    -0.000   -40.000    -0.000   720.000  3403.111  7200.000
6    -0.000    24.000    -0.000  -240.000    -0.000  7200.000 40837.337
illustrating the orthogonality of Hermite polynomials using GQ999IIb
         0         1         2         3         4         5         6
0     0.886     1.000    -0.000    -2.000     0.000    12.000    -0.000
1     1.000     1.772     2.000     0.000    -4.000     0.000    24.000
2    -0.000     2.000     7.090    12.000     0.000   -40.000     0.000
3    -2.000     0.000    12.000    42.539    72.000     0.000  -240.000
4     0.000    -4.000     0.000    72.000   340.311   720.000     0.000
5    12.000     0.000   -40.000     0.000   720.000  3403.111  7200.000
6    -0.000    24.000     0.000  -240.000     0.000  7200.000 40837.337
illustrating the orthogonality of Hermite polynomials using GQ4i
         0         1         2         3         4         5         6
0     0.299     0.286     0.285    -1.647     0.086     8.655     0.039
1     0.286     0.696     1.411    -0.033     0.350     0.322     0.434
2     0.285     1.411     1.644     1.196    -0.850     1.236     0.022
3    -1.647    -0.033     1.196     4.683     7.756    -1.042     8.101
4     0.086     0.350    -0.850     7.756    15.400    43.008     0.028
5     8.655     0.322     1.236    -1.042    43.008    55.947   344.797
6     0.039     0.434     0.022     8.101     0.028   344.797   215.718
illustrating the orthogonality of Hermite polynomials using GQ96i
         0         1         2         3         4         5         6
0     0.299     0.286     0.285    -1.647     0.086     8.655     0.039
1     0.286     0.696     1.411    -0.033     0.350     0.322     0.434
2     0.285     1.411     1.644     1.196    -0.850     1.236     0.022
```

[9] More on integration techniques can be found in my book entitled, *Numerical Calculus*, at https://www.amazon.com/dp/B07BS1DN1S

```
3   -1.647   -0.033    1.196    4.683    7.756   -1.042    8.101
4    0.086    0.350   -0.850    7.756   15.400   43.008    0.028
5    8.655    0.322    1.236   -1.042   43.008   55.947  344.903
6    0.039    0.434    0.022    8.101    0.028  344.903  215.718
creating surface: Hermite.tb2
  size: 128x128
```

The last of these three codes (fig2.c) contains the data and regression. The first program also spits out a file containing a surface composed of weighted Hermite polynomials. A complete Windows® program (with source code) to display a surface in 3D can be found in the folder examples\surface. If you drop a surface file (*.tb2) onto the executable (surface.exe), it will display. You can rotate and zoom the surface with the arrow keys, Home, End, PgUp, and PgDn.[10]

Contaminant in Groundwater

I was once involved with a large remediation project to quantify and clean up a certain contaminant in groundwater. This particular situation lends itself to one-dimensional analysis because of the terrain, subsurface conditions, and prevailing flow. We know the contaminant spread out somewhat radially and ends at some distance from the likely source location. We select a characteristic radius to roughly match the overall reduction in concentration, which turns out to be about one-third of the distance from the source so that the normalized domain extends from about -3 to +3. Soil samples were collected every 20 meters. Approximations of order 5, 10, 15, 20, 25, 30, and 35 are shown in the following figure:

[10] More on 3D rendering can be found in my book entitled, *3D Rendering in Windows®*, at http://www.amazon.com/dp/B01KG97XB8

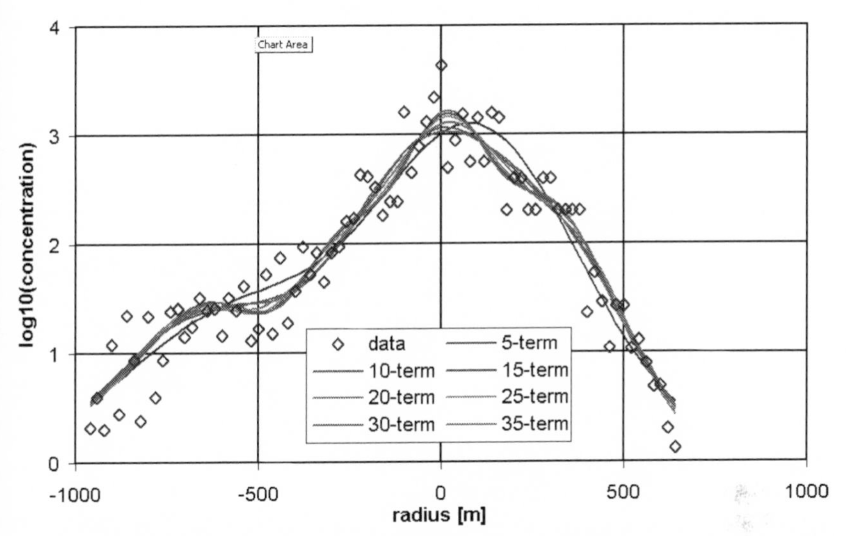

The 5-term approximation is probably adequate, but we go all the way up to 35 terms just to make a point: Conventional curve fitting (i.e., $y=a+bx+cx^2+\ldots$) goes berserk if you run up to 35 terms; whereas, orthogonal functions don't. This way, we don't have to worry if 10 is enough or 15 too many or if the approximation outside this interval will go wild. The code to pivot the data and calculate the coefficients with the weighting function are almost trivial:

```
for(i=0;i<m;i++)
  {
  for(c[i]=s=k=0;k<n;k++)
    {
    x=data[k].r/radius;
    y=data[k].c;
    w=exp(-x*x/2.);
    for(j=0;j<i;j++)
      y-=c[j]*Hermite(j,x)*w;
    h=Hermite(i,x);
    c[i]+=h*y*w;
    s+=h*h*w*w;
    }
  c[i]/=s;
  }
```

Evaluation is also a simple matter:

```
double Evaluate(double*c,int m,double r)
  {
  int i;
  double w,x,y;
  x=r/radius;
  w=exp(-x*x/2.);
```

```
for(y=i=0;i<m;i++)
  y+=c[i]*Hermite(i,x)*w;
return(y);
}
```

All the associated files are in the examples\Hermite folder.

Chapter 5. 2D Data Applications

We will now consider two-dimensional applications related to data analysis. We will consider images—a special type of 2D data—in the next chapter. The most obvious 2D data that everyone is familiar with is topography. We most often receive topographic data in the form of elevation contours, such as those shown in the following figure:

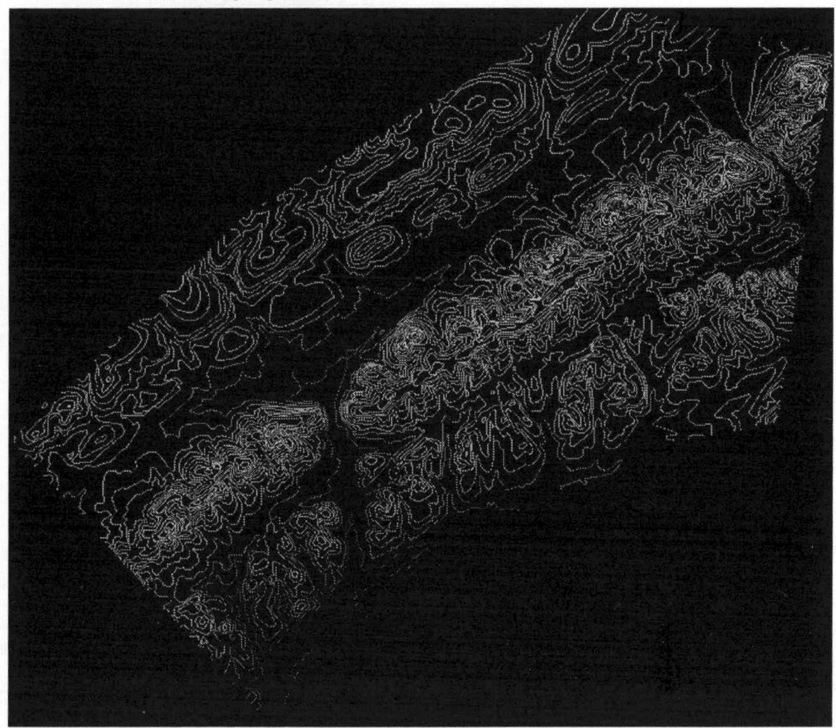

We will approach this data with conventional as well as orthogonal techniques. All of the associated files can be found in the examples\topography folder. The first thing we do is transform the coordinates by translating so that the center is at (0,0) and rotating so that the data lie along the principal axes (x and y). We find the rotation angle using Brent's algorithm to minimize the bounding rectangle. The transformed contours are shown in the figure on page ii, just before the table of contents. The transform 3D surface is shown just below that.

The conventional approach is to utilize the inverse distance method to interpolate the contours onto an evenly spaced grid, which can be slow even if you don't use atan2(y,x). Instead, you want to use the following simple code to determine which octant (or quadrant) you're in:

```
q=0;
```

```
if(dX>0.)
  q|=1;
if(dY>0.)
  q|=2;
if(fabs(dX)>fabs(dY))
  q|=4;
```

Calling atan2(y,x) over and over again will take much longer than the rest of the operations combined. There's no reason to weight the data, so we will use 2D Legendre polynomials to approximate and then interpolate the surface onto the same grid, comparing the time required as well as the quality of the result. Abbreviated program output is listed below (elapsed times are bold):

```
examples\topography>topography
reading topo: topography.p3d
  25379 lines read
  24797 points found
  291 polys found
  739943≤east≤745323, span=5380
  3976355≤north≤3981260, span=4905
  760≤elev≤1080, span=320
normalizing topo
  rotation=-32.8°
writing topo: transformed.p3d
inverse distance interpolation
  surface: 128x128
  137.260 seconds
writing surface: inverse_distance.tb2
Legendre approximation
  surface: 128x128
  polynomials: 16x16
  4.341 seconds
writing surface: Legendre16x16.tb2
Legendre approximation
  surface: 128x128
  polynomials: 32x32
  39.124 seconds
writing surface: Legendre32x32.tb2
Legendre approximation
  surface: 128x128
  polynomials: 48x48
  138.766 seconds
writing surface: Legendre48x48.tb2
Legendre approximation
  surface: 128x128
  polynomials: 64x64
  339.224 seconds
writing surface: Legendre64x64.tb2
```

Results in 3D for the 16x16 approximation are shown in this next figure:

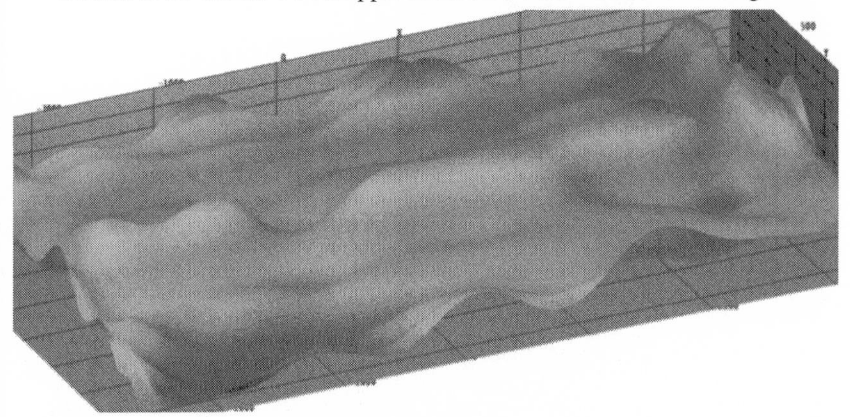

Not surprisingly for 16x16 polynomial approximation the results are quite different from that obtained with the inverse distance method. Which method you choose depends on what best suits your needs. We can increase the approximation order to add detail. The 64x64 approximation is shown below:

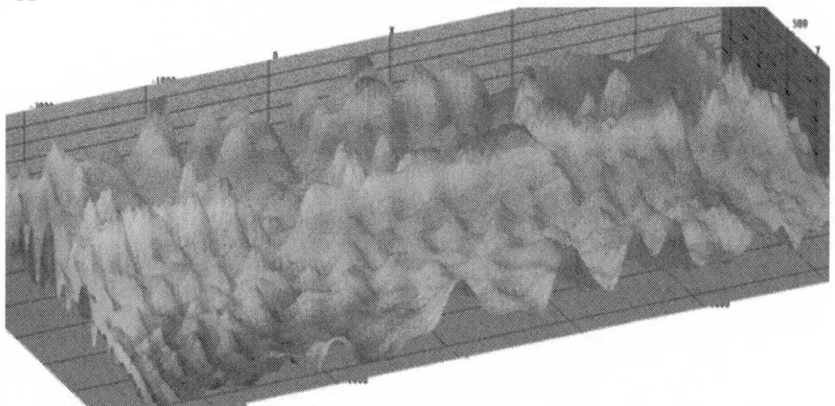

As with the 1D cases, we can continue increasing the order without encountering the same sort of problems associated with conventional curve fitting. The time consumed by the inverse distance method is directly proportional to the number of data points times the number of output points. If you want more spatial resolution, it will take longer. The Legendre method doesn't work this way. The vast majority of time is consumed calculating the coefficients. The time required for projecting without pivoting is proportional to the number of coefficients and the number of data points. The time required for pivoting goes up with the square of the number of coefficients because we must recalculate the polynomials each time (i.e., 1+2+3+...n=n*(n+1)/2). We can reduce this effort from n*(n+1)/2 to 2n by allocating add ional memory and saving the pivoted values.

More Contamination Data

Contaminant concentrations are often presented as two-dimensional scattered data. We had no reason to control the elevations at infinity in the previous example, but concentrations must vanish at extreme distances, so we switch from Legendre to Hermite polynomials for this next example. We also know that concentrations don't change abruptly in time or space, at least for contaminants dumped on the ground and left to seep into the soil.[11] Typical data for such lamentable conditions is illustrated in the next figure:

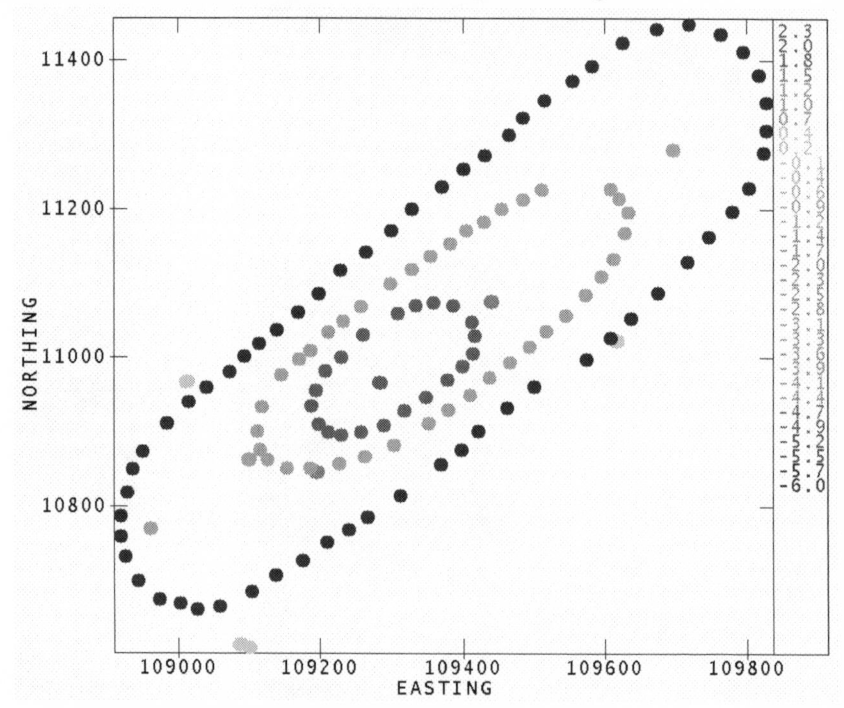

A plot similar to the preceding one has already been set up for each case (launch> TP2 *.TP2 to display all six).[12] The concentrations may be considered logarithmic, as in this case, or linearly, as in the other 5 cases provided. You will

---

[11] Let's not discuss where or what these are or who put them there so that I don't get sued.

[12] TP2 is my versatile plot program similar to Tecplot®, only it handles many more types of data sets, plus it's free. You can find it at:
http://dudleybenton.altervista.org/software/index.html

find 6 contaminant plumes in the folder examples\Hermite\2D. These are summarized in the following table.

name	preferred approximation
CSX	inverse distance
EBR	Hermite
FS2	Hermite
LF1	uncertain
WAF	Hermite
WAS	inverse distance

The program (plume2D.c) reads the data, transforms it, builds a surface using inverse distance interpolation, and then several orders of approximation using Hermite polynomials. You can drop a file on the .exe (*.p2d) or it will launch with the first one it finds in the folder. You can also plot the raw data by launching> TP2 *.p2d. A case where the inverse distance method works well (WAS plume) is shown in this next figure:

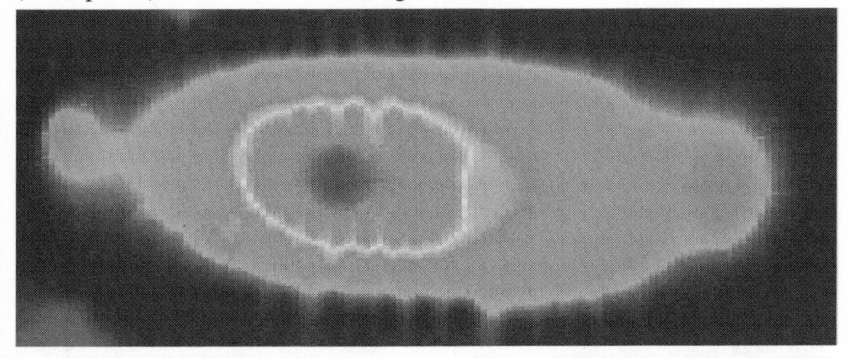

An example where the inverse distance is completely unsuitable (EBR plume) is shown next:

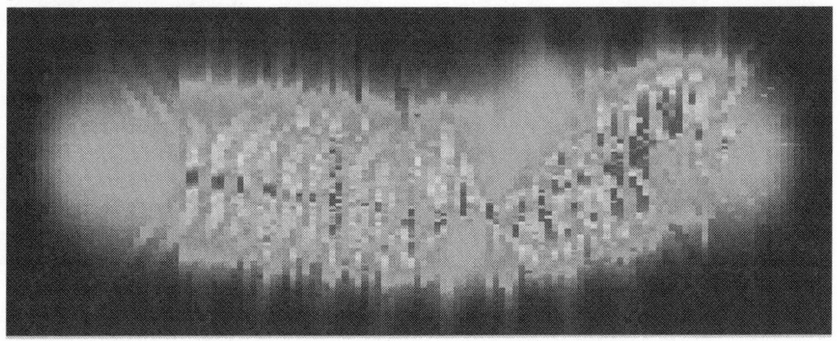

The WAS plume using Hermite approximation is shown in the next figure and has undesirable artifacts:

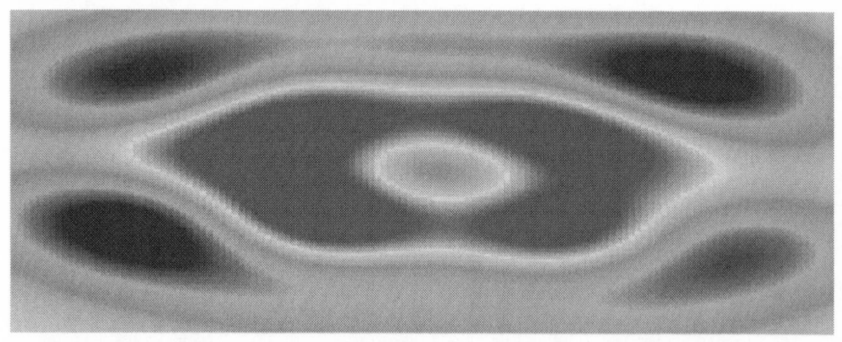

The EBR plume using Hermite approximation is quite satisfactory:

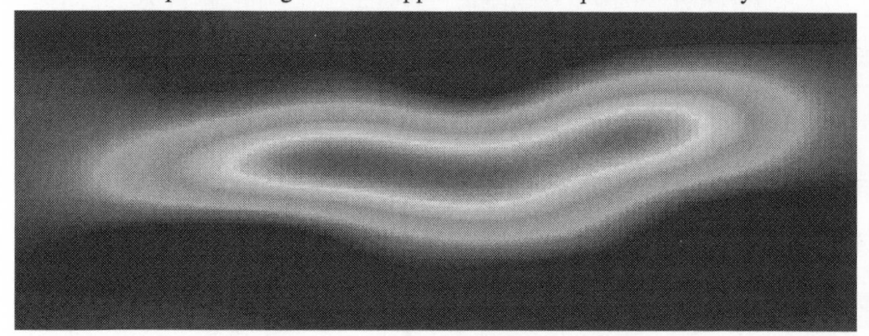

One that is questionable is the CSX plume with inverse distance. It is highly unlikely that the concentrations are so high over such a space and drop off so abruptly. The Hermite approximation if very blotchy and is also undesirable.

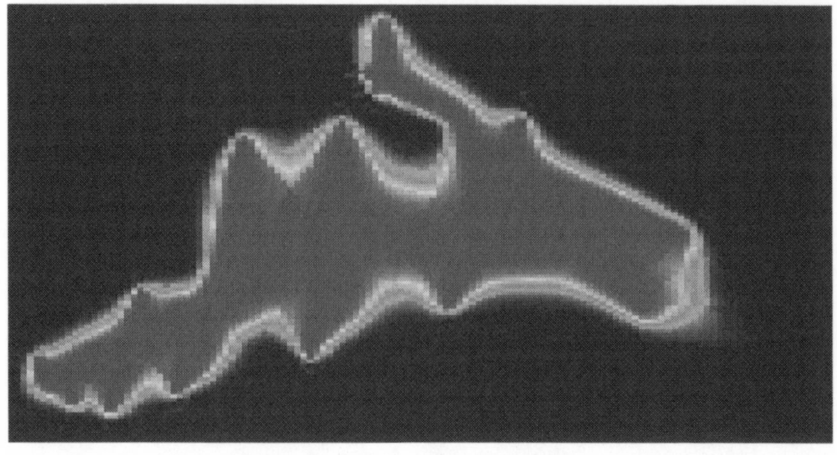

The WAF plume using inverse distance is also unacceptable:

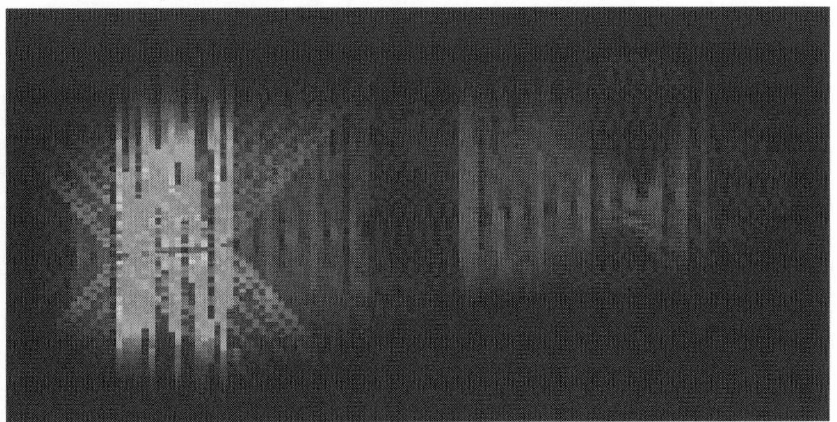

The Hermite approximation for the WAF plume is excellent:

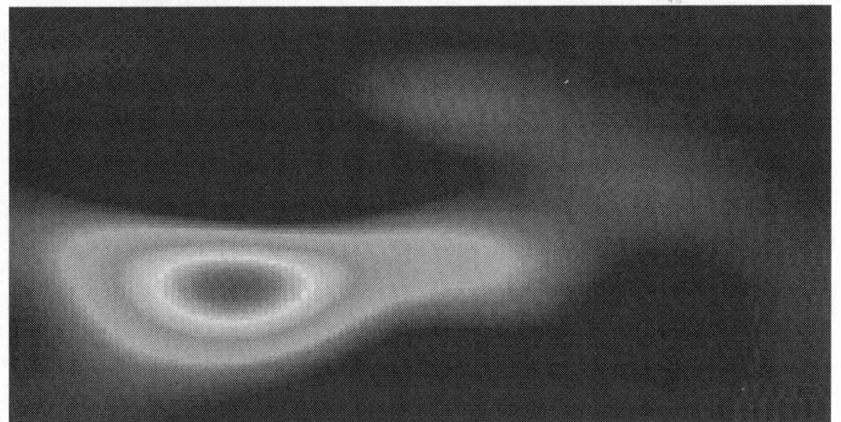

We will consider some of these same contaminant plumes in Chapter 7 when we get to 3D approximations.

Magnetometer Data

A magnetometer is often used to locate and classify unexploded ordinance (e.g., bombs, land mines, etc.). Data are collected by a technician—better still a robot—canvasing the area, paying particular attention to *hot spots*. The data points are scattered about and do have a pattern, but don't lend themselves to conventional plotting techniques. A clear graphical representation is important in planning the most efficient and safe extraction. This may also help with identifying the type and strength of device. The following figure is typical of such data:

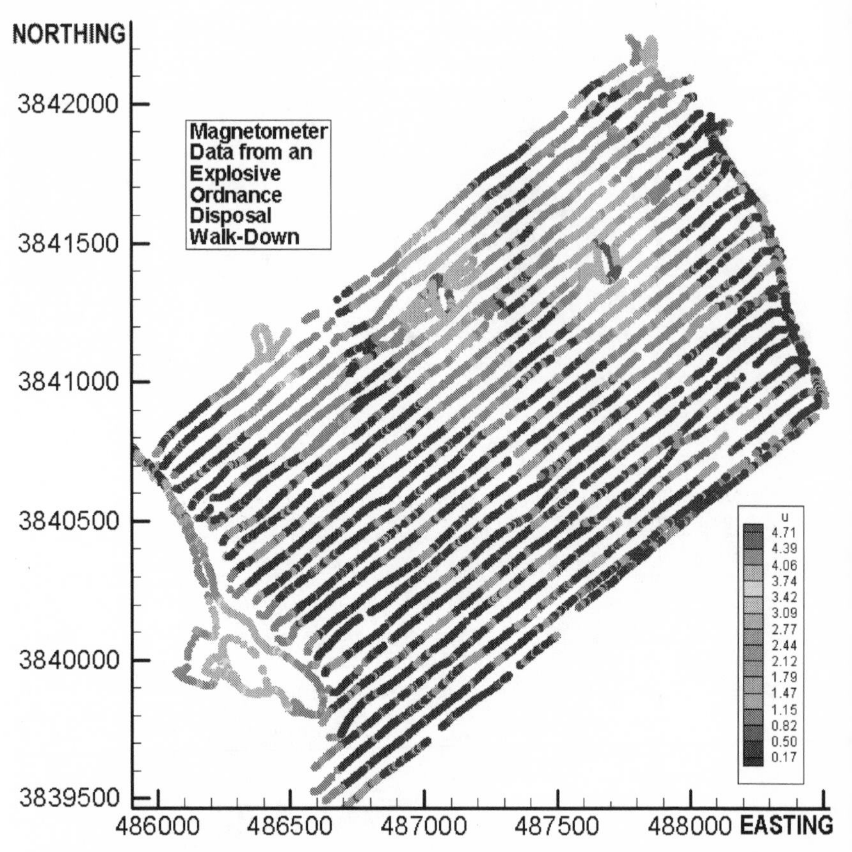

In this case perhaps five hot spots can be identified. We will use both inverse distance and Legendre polynomials as before to analyze and display the information. The data files and program code can be found in the folder examples\magnetometer. The magnetometer.c code is very similar to the topography.c program, only optimized to process randomly scattered data rather than contours. The results and output are also similar. You can display these in 2D using TP2 (e.g., TP2 inverse_distance.tb2 -flat) or 3D using TP2 (e.g., TP2 inverse_distance.tb2) or in 3D using the program in examples\surface (e.g., drop the xxx.tb2 file onto surface.exe). The inverse distance method in 3D is shown in the following figure, displayed by surface.exe:

The 16x16 Legendre approximation is:

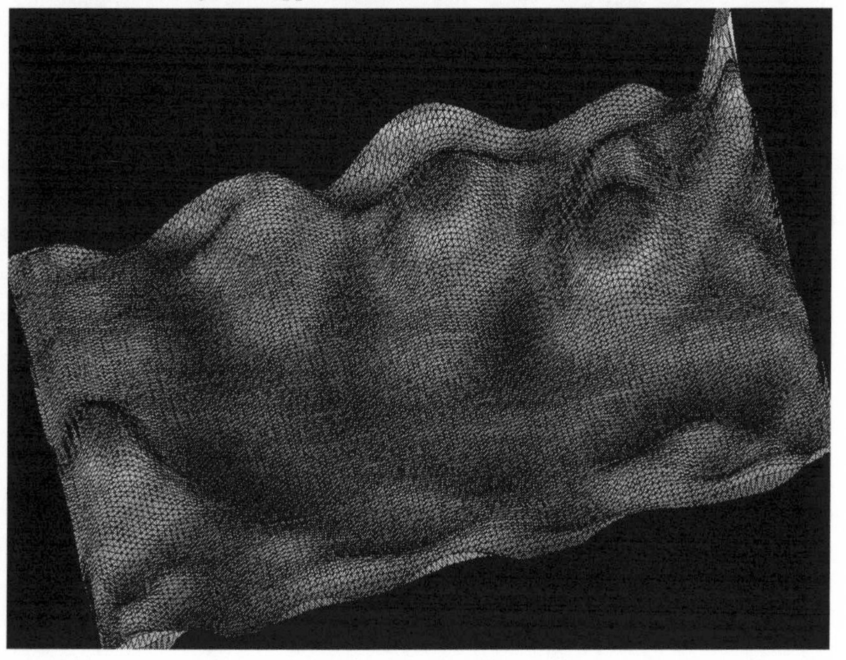

32x32, 48x48, and 64x64 approximations are also calculated. The 64x64 looks quite similar to the inverse distance. Both contain far more noise than is desirable for this application. This last surface reveals one large and five or six medium sized objects plus a half-dozen small objects.

Bathymetry Data

Bathymetry data is like topography contours in that it typically comes from a boat equipped with side-scanning sonar and follows a circuitous path. It's like the magnetometer data in that it varies along the path rather than being constant, as in the case of elevations. The following is typical of such data:

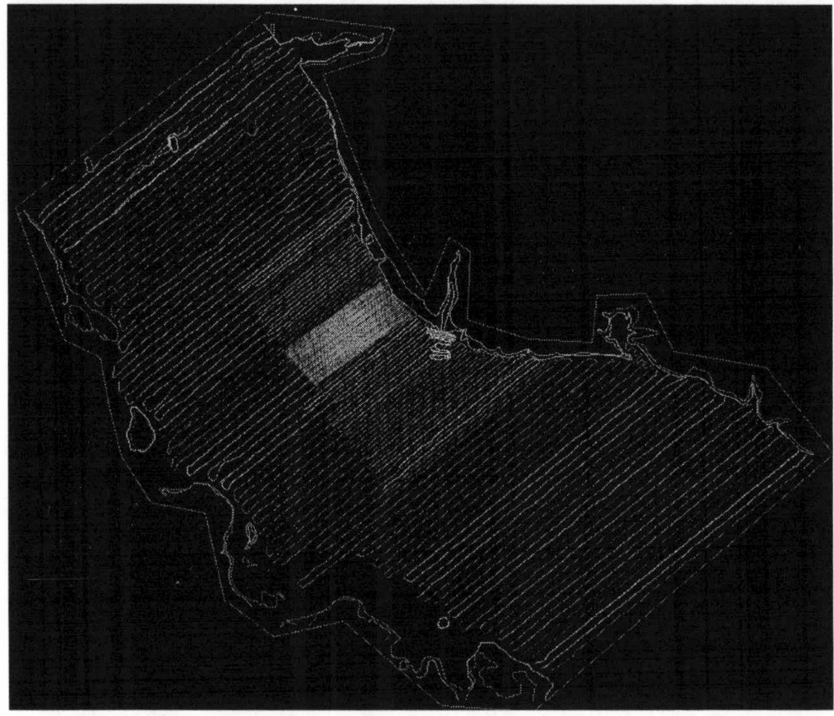

Some of the contours are connected and some are not. There are more contours in areas of particular interest. In this case there are 36,700 data points, making the direct inverse distance method quite burdensome. TP2 has the capability of grouping scattered data to narrow and speed up the search, but this is a programming strategy, unrelated to orthogonal functions, so we will not consider it here. The magnetometer code requires only minor modification and can be found in the examples\bathymetry folder. Once again, the inverse distance method is too bumpy, but the Legendre is smooth. Results of the inverse distance method are shown in this first figure and for the 16x16 Legendre in the second.

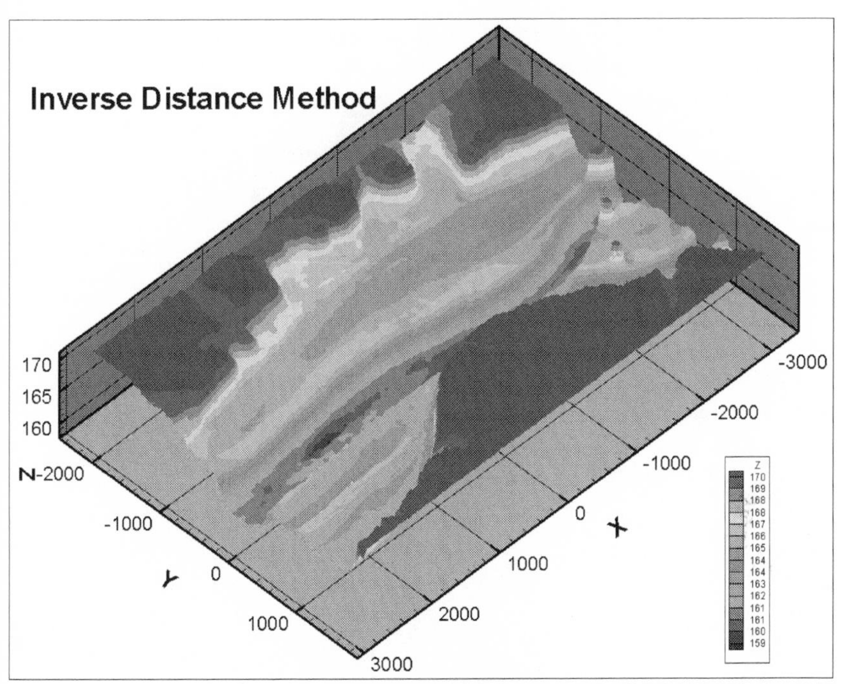
Inverse Distance Method
170
165
160
N-2000
-1000
Y
0
1000
3000
2000
1000
0
X
-1000
-2000
-3000
Z
170
169
168
168
167
166
165
164
164
163
162
161
161
160
159

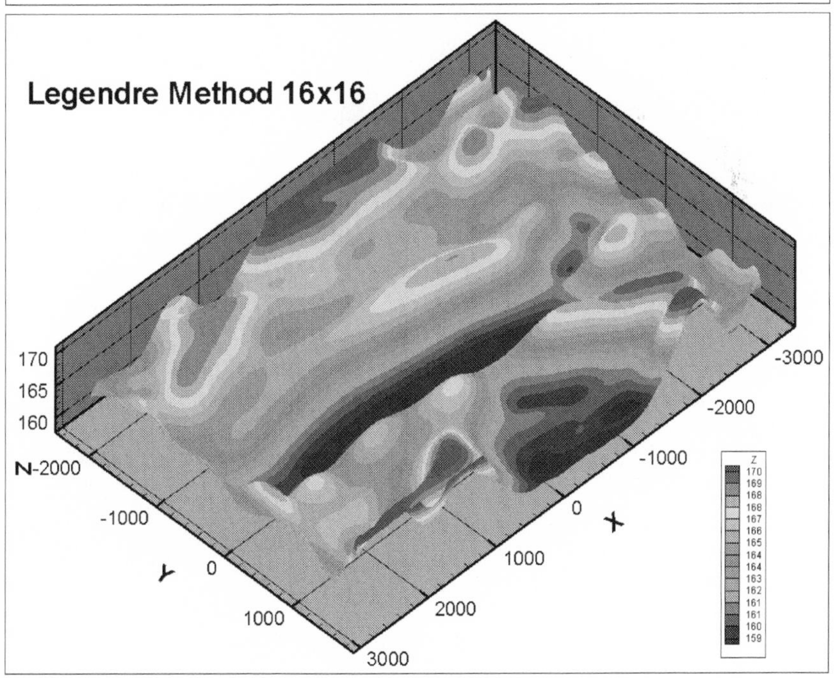
Legendre Method 16x16
170
165
160
N-2000
-1000
Y
0
1000
3000
2000
1000
0
X
-1000
-2000
-3000
Z
170
169
168
168
167
166
165
164
164
163
162
161
161
160
159

There is an overbank artifact in the second figure but not the first. This arises from a lack of data in the region shown below:

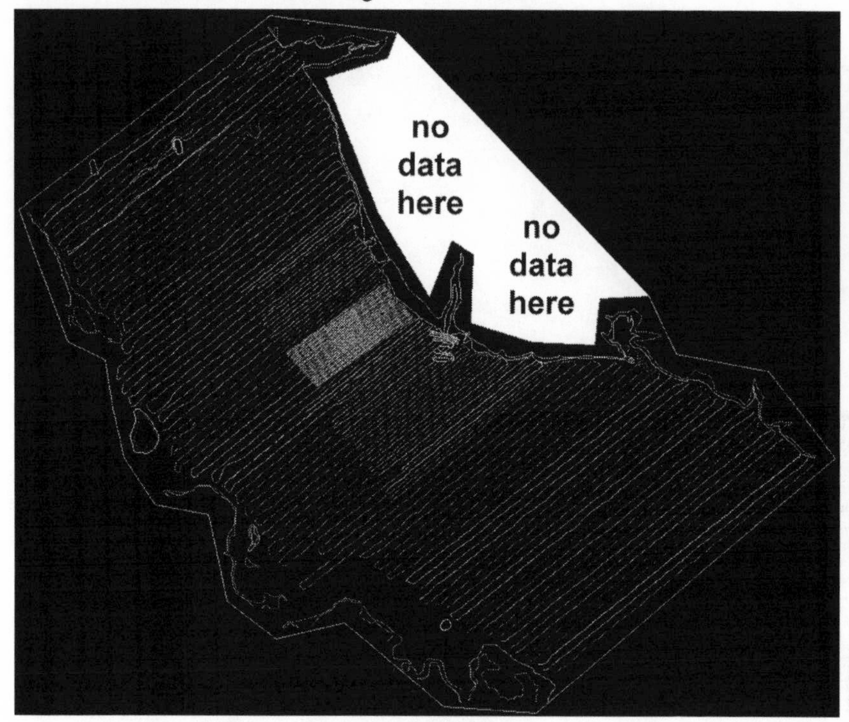

The inverse distance method works the way we want it to in this case by continuing the elevations flat into the zone without data; whereas, the Legendre method doesn't. We can easily fix this by sprinkling data points with constant elevation in this region. After adding these points, the 16x16 Legendre method results are satisfactory, as illustrated in the following figure:

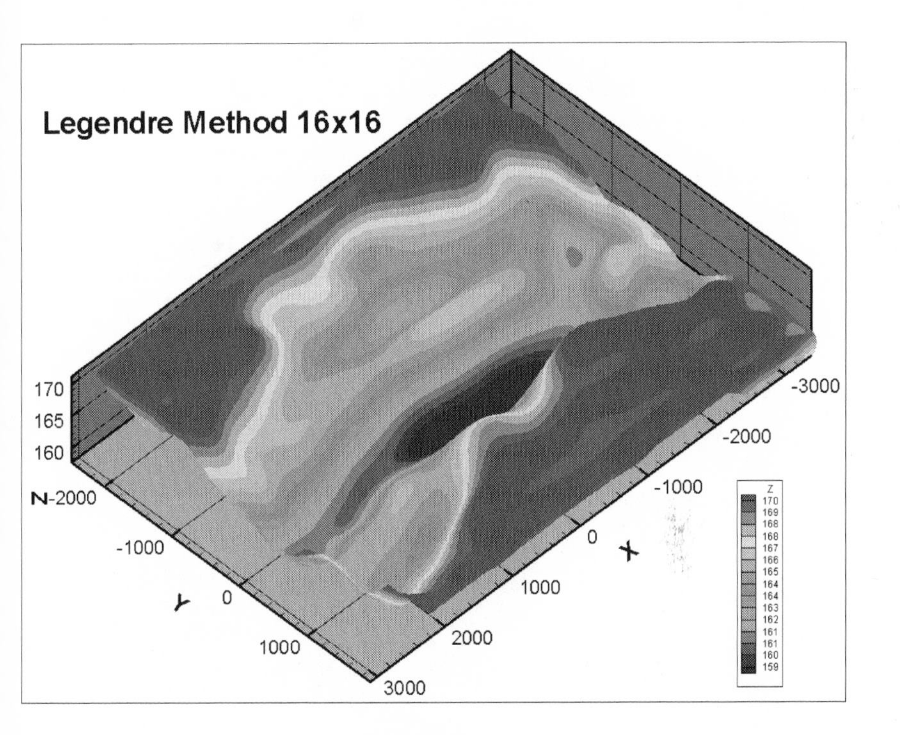

<u>Solar Data</u>

We will next consider solar data collected over an array by 284 instruments. Direct Normal Insolation (DNI) is recorded every minute throughout the day and night. We will use the location of the sensors along with the readings and Legendre polynomials to paint a map of intensity for each minute and write these out as individual BMP files. They can be gathered together into a GIF. We will embed the time stamp in each frame. The BMP to GIF tool (bmp2gif.c) can be found in the examples\utilities folder. If you want to skip the BMP step and go straight to the GIF, refer to the next section. There's also a GIF to BMP tool (gif2bmp.c) in the utilities folder.

The data and associated files can be found in the folder examples\solar. We read the data one line of data at a time, creating an image if the average DNI meets some criteria (at the top of the code), as there's no point creating a solar map in the dark or on a completely clear day. The selection criteria are entirely arbitrary and intended to isolate interesting times.

```
/* DNI criteria for generating an image */
double min_must_be_no_more_than=500.;
double max_must_be_no_less_than=900.;
double std_must_be_at_least    = 50.;
double expected_clear_sky_peak=1075.;
```

The spatial map is colored blue to green to yellow to orange to red to magenta, based on the interpolated intensity, which is derived from the 284 instruments. The instrument locations are contained in a data statement. You could adapt the code to handle some other spatially varying time series data or instrument arrangement.

```
struct{char*name;double x,y,dni;}Sensor[]={
  {"101A", 0,36},{"101B", 0,32},{"102A", 2,36},
  {"103B", 4,32},{"104A", 6,36},{"104B", 6,32},
  etc...
  {"441A",92, 4},{"441B",92, 0},{"442A",96, 4}};
int sensors=sizeof(Sensor)/sizeof(Sensor[0]);
```

One of the more interesting frames is shown below:

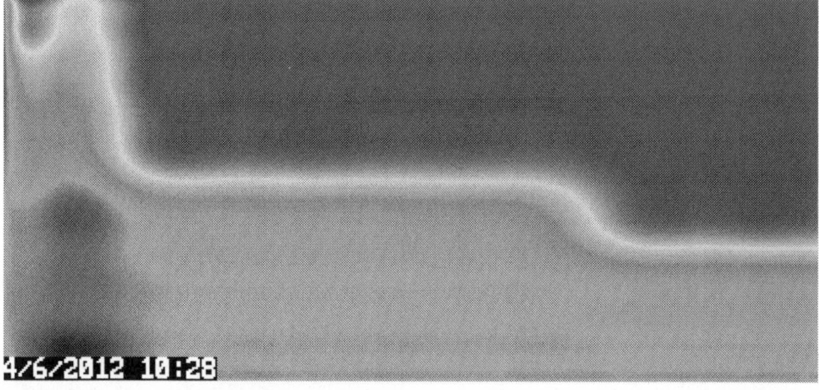

Another interesting one is:

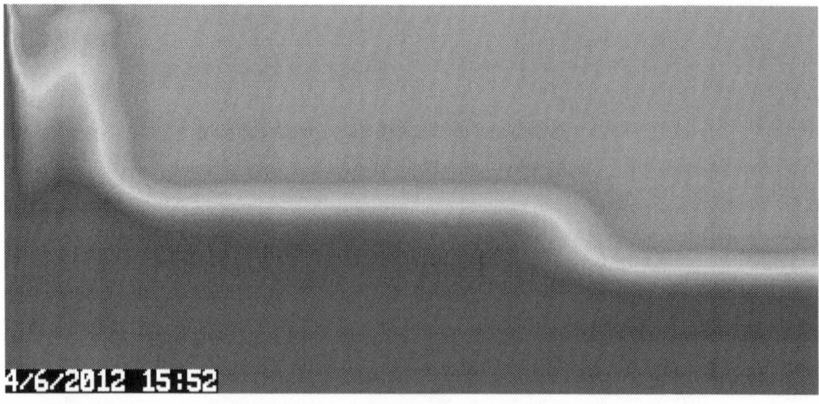

The image resolution is also calculated to roughly match the aspect ratio of the instrument array. You could easily adjust this:

```
wide=GetSystemMetrics(SM_CXSCREEN)/3;
high=GetSystemMetrics(SM_CYSCREEN)/3;
```

```
if(wide/(Xx-Xm)<high/(Yx-Ym))
  {
  bi.biWidth=wide;
  bi.biHeight=(unsigned)((Yx-Ym)*wide/(Xx-Xm));
  }
else
  {
  bi.biHeight=high;
  bi.biWidth=(unsigned)((Xx-Xm)*high/(Yx-Ym));
  }
```

You can also easily change the embedded text and its location:

```
sprintf(bufr,"%i/%i/%i %i:%02i",mo,da,yr,hr,mn);
EmbedText(0,FontHigh-1,bufr,white,black,FALSE);
```

Contrary to popular assumption, solar irradiance is non-uniform and changes throughout the day. Haze and clouds significantly impact the power input and, thus, the power output. A clear day might look like this:

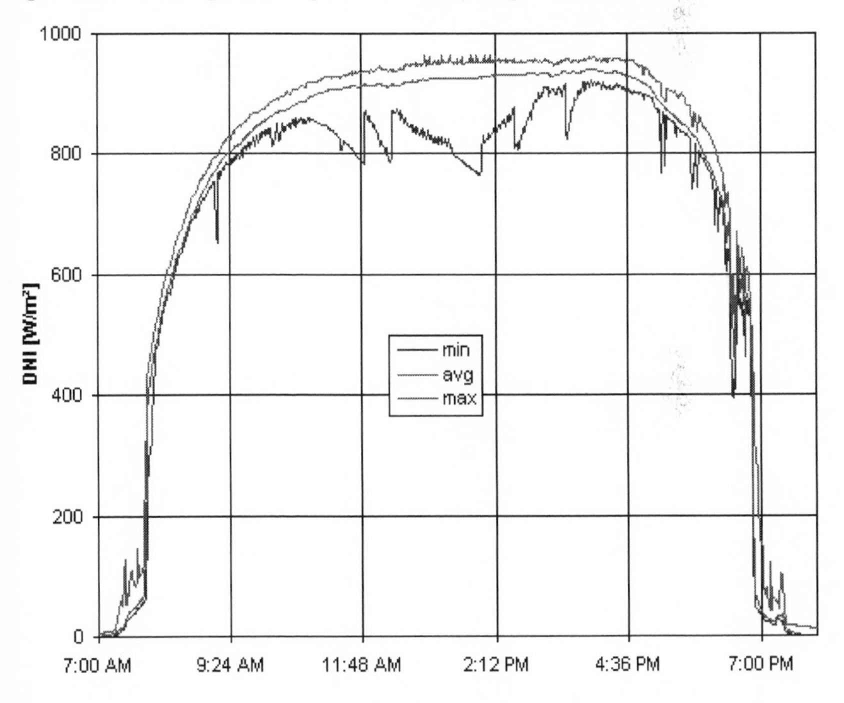

A cloudy day looks more like this:

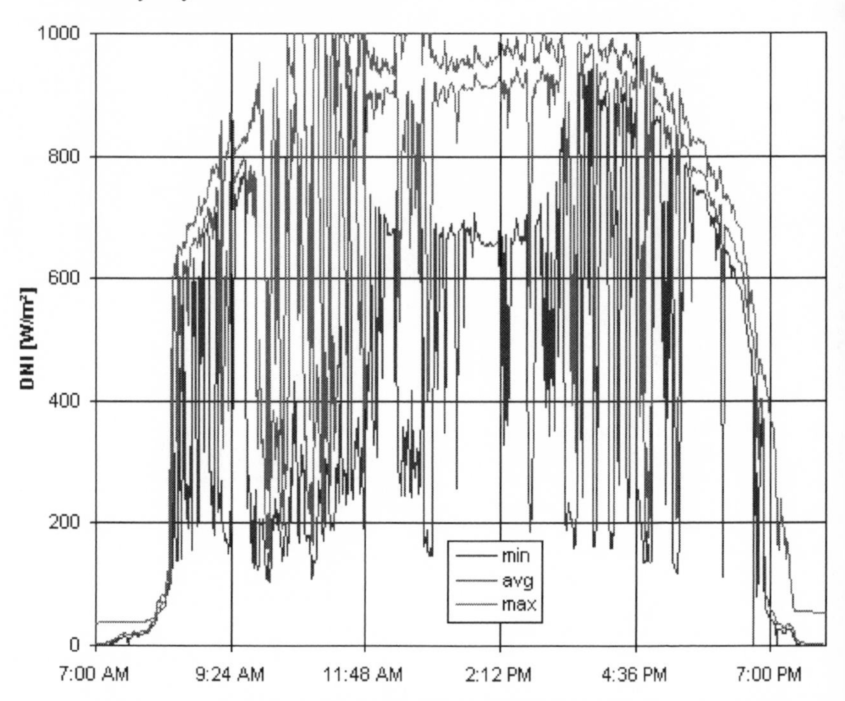

Everyone presumes that Phoenix, AZ is clear most of the time, but this is far from true, as shown in the following figure:

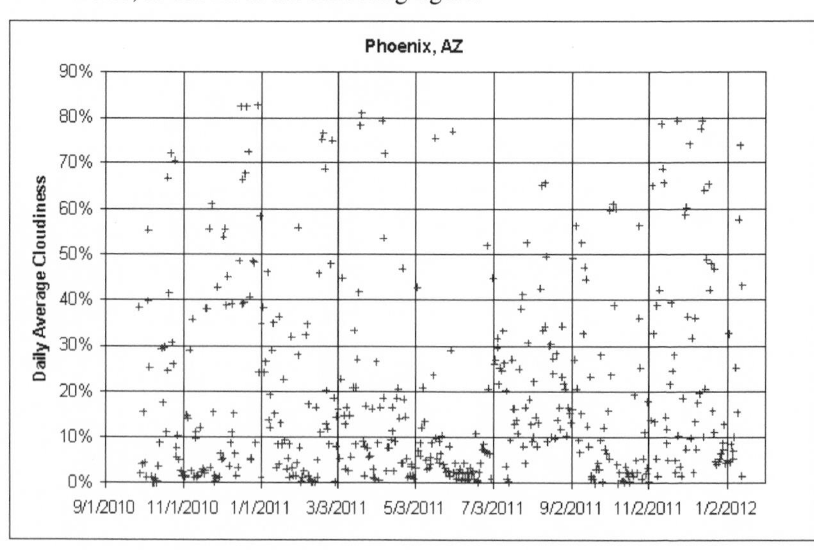

The actual occurrence of clear days is much less than you might think:

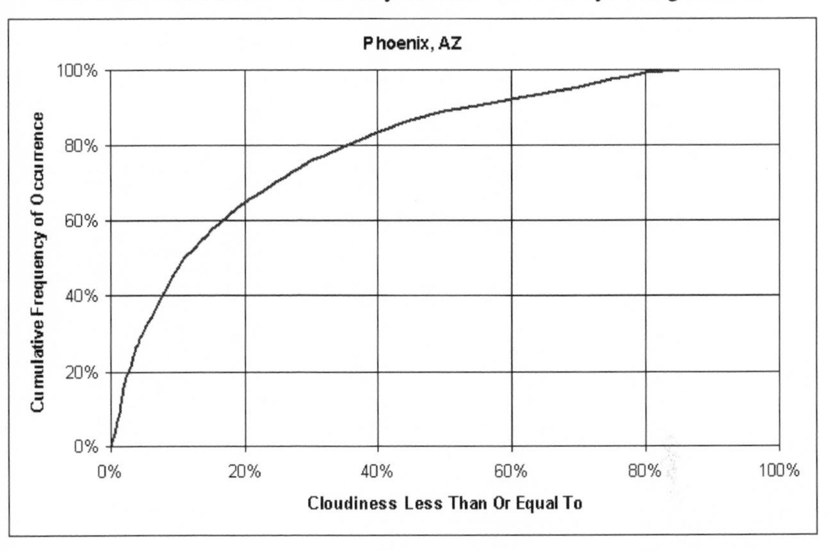

Air Quality Data

Solar incidence doesn't fade away with distance, so we used Lagrange polynomials in the preceding section. Air pollutants—ozone in particular—does fade away over time and diminish with distance from the source.

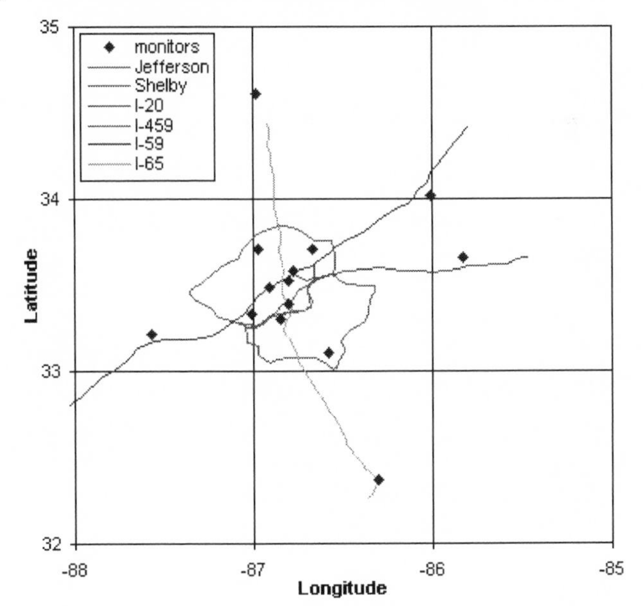

Ozone is produced by motor vehicles and some industrial processes. You will find some interesting ozone data in the folder examples\ozone. The procedure is similar to the previous, only we will use Hermite polynomials. In addition to the time stamp we will also overlay a map of interstates and monitoring stations shown in the previous figure. The time series data at the monitoring stations is illustrated below:

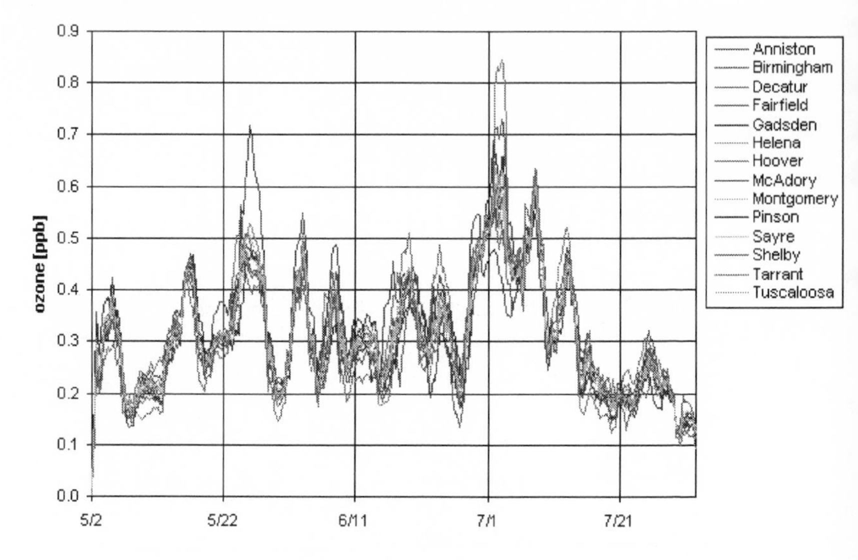

The following was a fairly mild day for ozone:

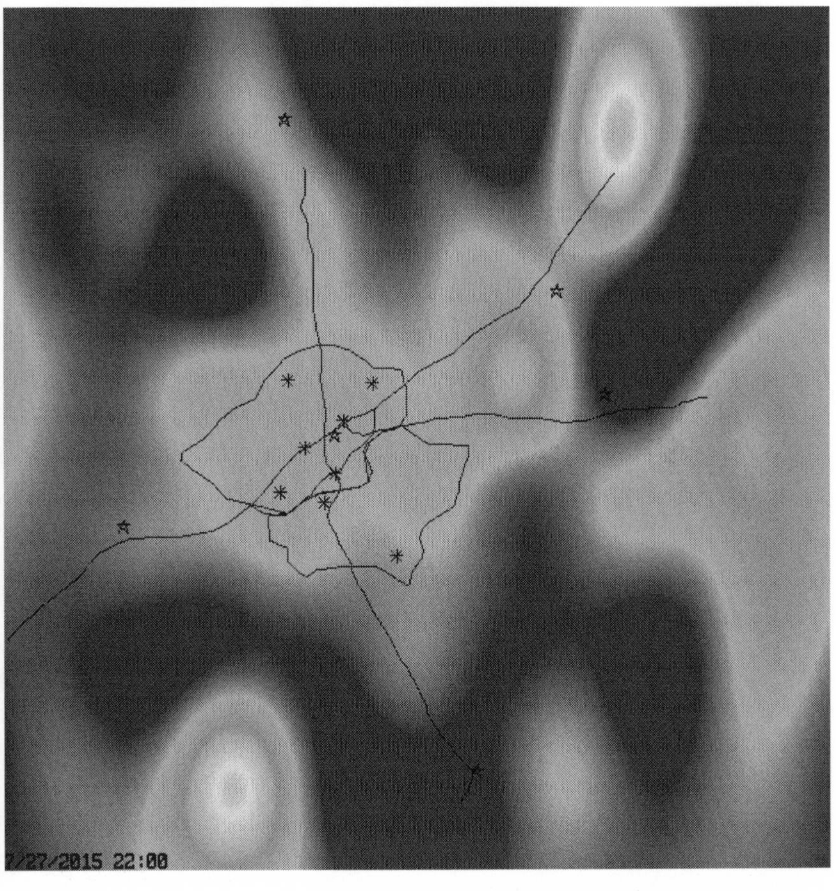

whereas, this next frame corresponds to a severe day:

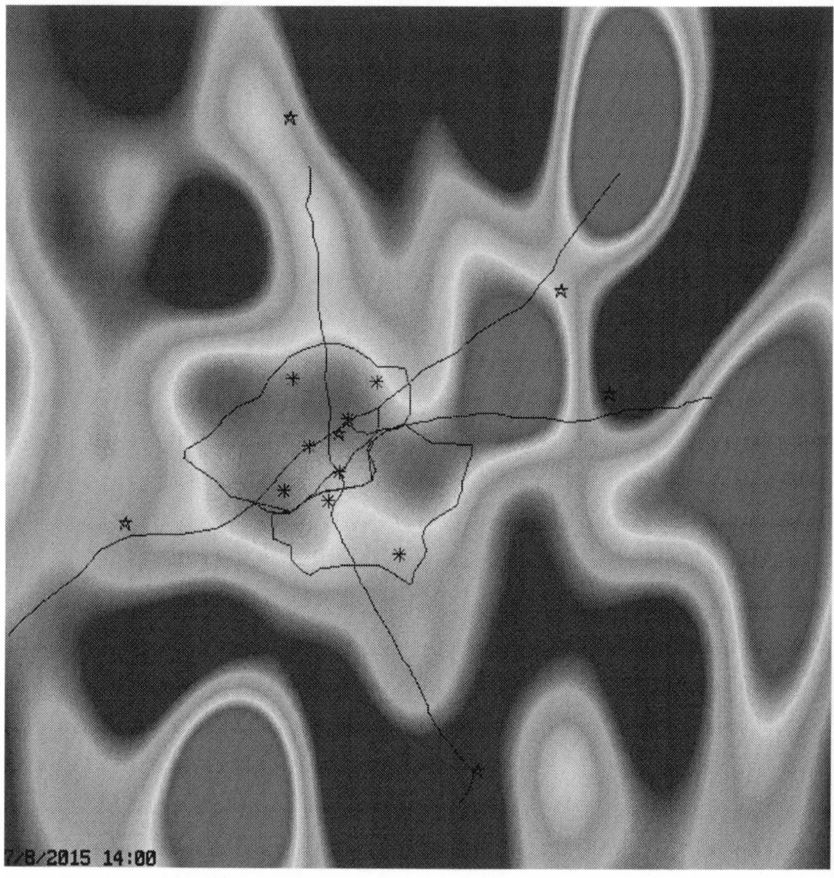

The entire 276-frame animation is in the ozone folder.

Chapter 6. Image Decomposition

There are many articles on the Web related to orthogonal functions and image processing. Like the Radon Transform, one must ask, "Why?" If the purpose is to store or transmit an image over a network or the Web, that's not a problem. If you want preserve or transmit the image precisely, use a GIF or TIF. If you don't mind losing some of the information, use a JPG. A photograph should be stored as a JPG; whereas, a line drawing or anything with text should be stored as a GIF or TIF. Never store a line drawing or clipart as a JPG and never store a photograph as a GIF. Sometimes Acrobat® uses the wrong compression algorithm for charts and graphs (storing them as a JPG) and pictures (storing them as a GIF). This mistake in settings results in very poor quality documents.

The JPG compression algorithm uses a Fourier Transform at one stage, but this is not the primary focus of the process and will not be included in our discussion here. The reason one considers the Radon Transform is that's the way the scanner or seismograph spits out the data. We don't use the Radon Transform because of its efficiency or convenience; rather we use it by necessity. If we want to transmit images back from Voyager across a billion km, we would have reason to consider compression algorithms in a much different light than conserving disk space or network bandwidth. Before we get into the images, there is some math to cover first.

Singular Value Decomposition of a Matrix

There are many ways of analyzing matrices. One of the most interesting is the singular value decomposition. Any matrix can be represented as a product of three matrices, each with special characteristics. These are the left and right eigenvectors (*[U]* and *[V]*) and the singular values (*[S]*). The form is:

$$[A]=[U]^T[S][V] \tag{6.1}$$

The eigenvectors are orthonormal, that is, they are both orthogonal and have a magnitude of unity (i.e., sum of the squares equals one). The orthonormal conditions means that:

$$[U]^T[U]=[V]^T[V]=[I] \tag{6.2}$$

where *[I]* is the identify matrix (i.e., ones on the diagonal and zeroes elsewhere). If *[A]* is symmetric, the left and right eigenvectors are complementary, $[V]=[U]^T$. (i.e., *[V]* is the transpose of *[U]*). The matrix *[S]* is diagonal (i.e. like the identity matrix, non-zero on the diagonal and zeroes elsewhere). If *[A]* is also positive definite, all of the eigenvalues are greater than zero. If *[A]* is singular (i.e., degenerate), one or more of the eigenvalues are zero. The number of rows and columns in *[A]* need not be equal.

If we were considering nodes along a vibrating beam, the matrix relating the stiffness and displacement could be expressed in terms of a singular value decomposition. The eigenvectors would be the modal shapes, which are sin(x), sin(2x), sin(3x), and/or cos(x), cos(2x), cos(3x), etc., depending on the end conditions. The singular values would be the square of the resonant frequencies. This is how orthogonal functions enter the discussion: eigenvectors are the orthogonal functions and eigenvalues are the coefficients. Any matrix—including an image—can be represented by a singular value decomposition. The question becomes, "Is there something to be gained here?" The answer is, "Yes!"

Since the eigenvectors are orthonormal, the contribution from any trio (left eigenvector, right eigenvector, singular value) is no larger than the magnitude of the singular value. If we arrange the eigenvectors and singular values such that the largest one is at the top left and the smallest at the bottom right, we can readily see the relative importance (i.e., contribution to the whole) of each part. For many images, the size of the singular values falls off quickly so that most of the information is contained in the first few, while the rest are just noise. If we're transmitting an image back from Pluto, we don't have to send the whole thing, just the most significant content and we have a way of answering the age-old question, "How much is enough?"

You may not be familiar with the means of finding a singular value decomposition of a matrix, but it's not too complicated. If you need an exact result, there is a straightforward approach called Givens Rotations.[13] It works well enough for small matrices, but isn't the best approach for larger ones, especially images. The Givens approach was replaced by a more efficient method called Householder Transformations.[14] You will find the code in folder examples\SVD. For illustration, we will consider the following simple matrix:

[A]=

1	1	1	1
1	2	4	8
1	3	9	27
1	4	16	64

You may recognize this as the left hand side of $y=a+bx+cx^2+dx^3$, $x=1, 2, 3, 4$. This matrix containing a geometric progression is called a Vandermonde.[15]

[13] Named after Wallace Givens, who developed this technique in the early 1950s.
[14] Named after Alston Scott Householder, who introduced this technique in 1958.
[15] Named after the French mathematician, Alexandre-Théophile Vandermonde.

The singular value decomposition is:

[U]=

-0.0199	-0.0701	-0.2567	-0.9637
0.3438	0.5523	0.7208	-0.2393
-0.7880	-0.2667	0.5440	-0.1092
0.5103	-0.7867	0.3444	-0.0451

[S]=

72.55	0	0	0
0	3.655	0	0
0	0	0.7303	0
0	0	0	0.06196

[V]=

-0.0181	-0.1226	-0.3937	-0.9109
0.3769	0.6613	0.5546	-0.3362
-0.8489	-0.0262	0.4917	-0.1921
0.3702	-0.7395	0.5438	-0.1428

The equivalent matrix [A] can easily be generated for any order. The program SVDtest.c performs this as a loop that you can adjust. The singular values for orders 2 through 8 are:

2.61803,0.381966

10.6496,1.2507,0.150156

72.5541,3.65504,0.730342,0.0619585

695.842,18.2499,2.00182,0.426073,0.0265896

8575.12,140.99,7.58787,1.32258,0.242885,0.0117274

128993,1465.27,48.1795,4.07498,0.93089,0.136586,0.00527384

2290650,19109.4,430.864,21.4119,2.57892,0.65721,0.0761592,0.00240587

Notice how rapidly the singular values fall off. Because the magnitude of the eigenvectors is always unity, this means that most of the information in these matrices is contained in the first few terms. Since we only need the first few terms, it's not necessary expend the computational effort to find them all. There is a much faster way of finding the largest eigenvalues, which are the only ones we're interested in. This technique is called the *power method*. It is also contained in the SVDtest.c file. This little program finds and lists the eigenvalues and eigenvectors as well as demonstrating that these actually work by calculating $[U]^T[U]$, $[V]^T[V]$, and $[U]^T[S][V]$ to arrive back at $[A]$. The following is an example of the output:

```
singular value decomposition
creating 3x3 Vandermonde
[a]=1 1 1
    1 2 4
    1 3 9
using Householder Transformations
[u]=-0.132386 -0.426381 -0.894803
    0.80141 0.485188 -0.349764
    -0.583281 0.763408 -0.277474
[s]=10.6496 1.2507 0.150156
[v]=-0.136491 -0.344574 -0.928784
    0.749045 0.57767 -0.32439
    -0.648306 0.739977 -0.179255
[uT][u]=1 0 0
        0 1 0
        0 0 1
[vT][v]=1 0 0
        0 1 0
        0 0 1
[uT][s][v]=1 1 1
           1 2 4
           1 3 9
```

The Power Method sounds good in theory, but there are a few problems, including: 1) knowing when it's converged, 2) even when it's converged to machine precision it's often still not close enough, 3) the eigenvectors aren't really orthogonal, 4) it produces the right eigenvectors but not the left and we're often dealing with asymmetric matrices, so these aren't the same, etc. There's not much that can be done to improve the estimated eigenvalues. The eigenvectors can be sequentially orthogonalized using the modified Gram-Schmidt procedure, but this can be problematic. The orthogonalization procedure is handled by:

```
for(i=0;i<k;i++)
  {
  p=dotproduct(v+cols*k,v+cols*i,cols);
  for(j=0;j<cols;j++)
    v[cols*k+j]-=p*v[cols*i+j];
  normalize(v+cols*k,cols);
  }
```

The dot product and normalization functions are simple:

```
double dotproduct(double*u,double*v,int n)
  {
  int i;
  double p;
  for(p=i=0;i<n;i++)
    p+=u[i]*v[i];
```

```
  return(p);
  }
void normalize(double*v,int n)
  {
  int i;
  double vv;
  vv=l2norm(v,n);
  for(i=0;i<n;i++)
    v[i]/=vv;
 }
```

The left eigenvalues can be estimated from the right ones and then orthogonalized in the same way.

```
    multiply(a,v+cols*k,u+rows*k,rows,cols,1);
    normalize(u+rows*k,rows);
    for(i=0;i<k;i++)
      {
      p=dotproduct(u+rows*k,u+rows*i,rows);
      for(j=0;j<rows;j++)
        u[rows*k+j]-=p*u[rows*i+j];
      normalize(u+rows*k,rows);
      }
```

All of these patches are included in SVDtest.c, which yields the following for the Power Method applied to the same problem as before:

```
using Power Method
[u]=0.137729 0.807899 0.573
    0.430706 0.472101 -0.769164
    0.891921 -0.352731 0.282945
[s]=10.6031 1.24544 0.151452
[v]=0.137729 0.430706 0.891921
    0.807899 0.472101 -0.352731
    0.573 -0.769164 0.282945
[uT][u]=1  0  0
        0  1  0
        0  0  1
[vT][v]=1  0  0
        0  1  0
        0  0  1
[uT][s][v]=0.711908 0.778327 1.15153
           1.62423 4.00821 7.4179
           0.0874157 2.13158 5.76896
```

Note the discrepancy in eigenvalues: 10.6031 (10.6496), 1.24544 (1.2507), 0.151452 (0.150156). Also the discrepancy in the last row of the reconstructed matrix: 0.0874157 (was 1), 2.13158 (was 3), 5.76896 (was 9). It's this bad for a 3x3 matrix, you can imagine the error for a 4096x4096 or larger one. The Power Method takes longer than Householder if you require more than a few eigenvalues and eigenvectors. Remember, we're not using this procedure because it's so convenient or so accurate or so efficient. We use it by necessity.

In this same folder you will find another program, ImageApprox.c, which applies this procedure to images. It currently only handles 8-bits/pixel gray-scale BMP images. You can easily adapt it to accept 24-bits/pixel color images as well as GIF and JPG formats using the code in the examples\utilities folder. Simply add the appropriate #include statements in the source and gif89.c jpeg6b.c in _compile_ImageApprox.bat, then recompile. We will first use the Householder transformation, using only some of the eigenvalues and vectors.

This first reconstruction uses only the first 4 orders:

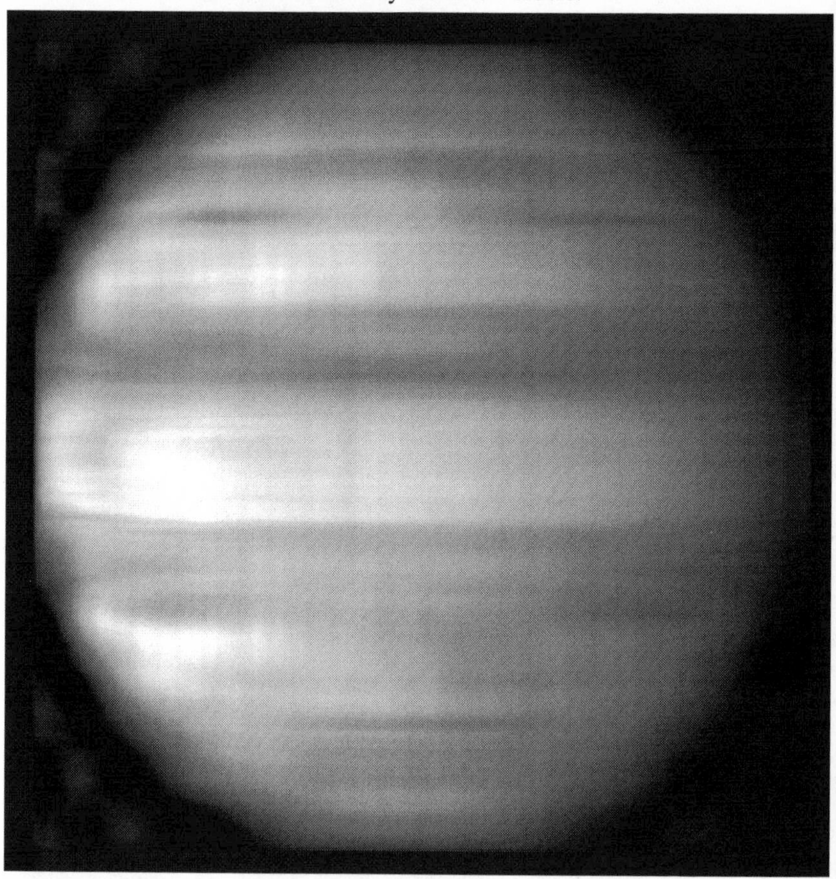

This isn't too bad, but that's deceiving. This reconstruction is only this good because the picture is symmetric and simple.

This next reconstruction uses the first 16 orders:

After 128 terms the image is almost indistinguishable from the original:

This is not surprising, considering how fast the eigenvalues fall off, as shown in this next figure on a log scale:

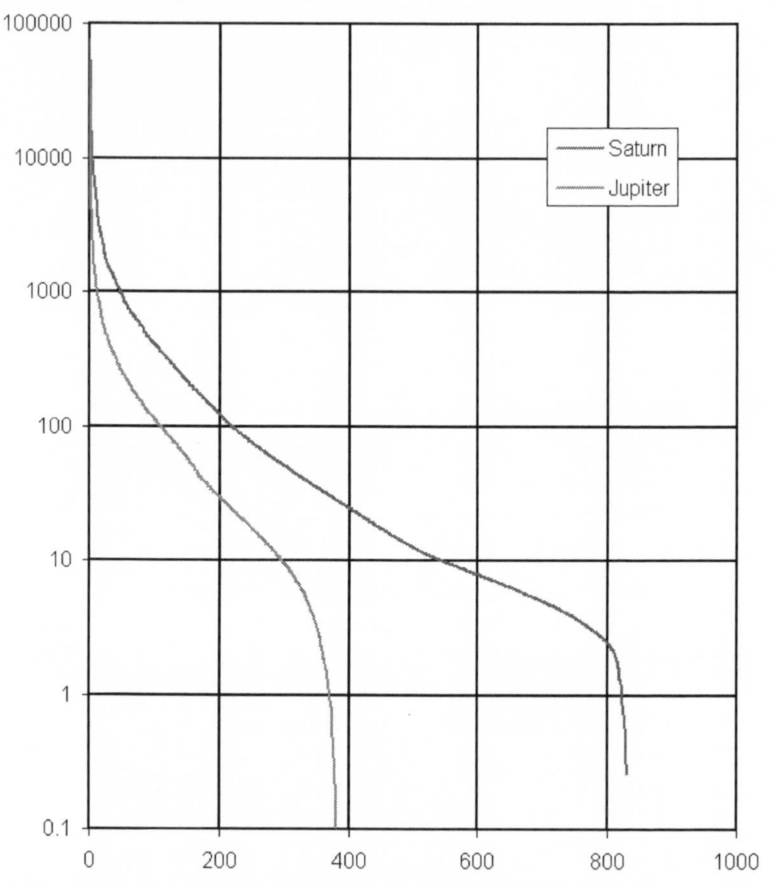

The first 10 eigenvalues/vectors contain the following cumulative information compared to the whole matrix:

1	44%
2	52%
3	56%
4	59%
5	61%
6	63%
7	65%
8	66%
9	67%
10	69%

ImageApprox.c also contains the Power Method, but this rarely produces acceptable results and it often takes longer. This next image is not as symmetric.

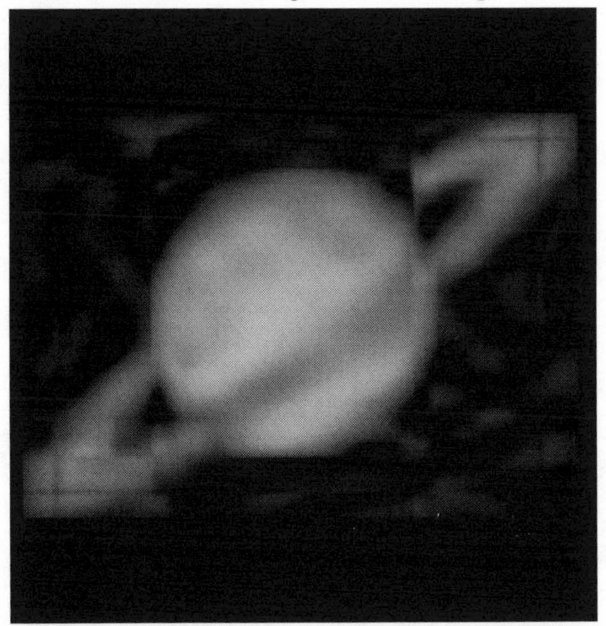

The 24th order image above is poor. The 48th is only slightly better:

The 128th order approximation is quite acceptable, even with only 10% of the original information:

This application is interesting from an historical perspective, but is of little practical value at the present. Contemporary compression algorithms first break the image into smaller blocks and provide more than adequate space saving as well as excellent processing speeds. JPEG is one such method.

Chapter 7. 3D Data Applications

We will now consider applications having three spatial dimensions, that is, fields representing some property that varies with height, width, and depth. The first of these will be geological, more specifically, the varying hydraulic conductivity throughout a volume of ground beneath the surface. The data points are shown below as spheres, colored proportional to their value.

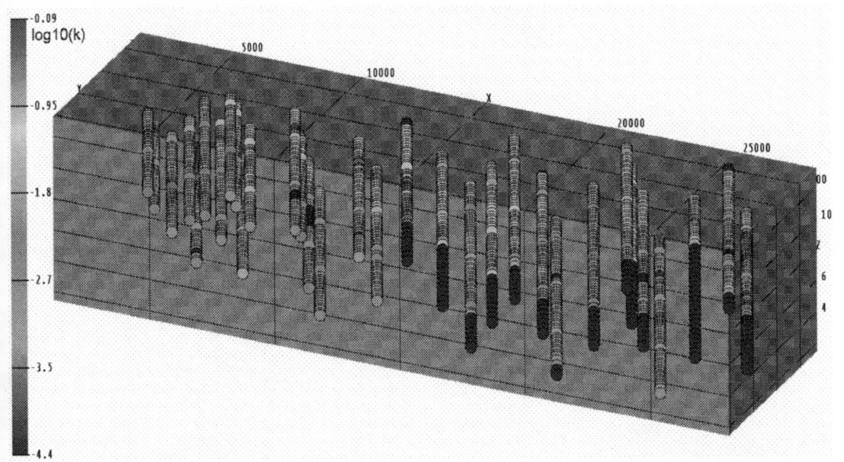

This view is from below, looking up toward the surface. The blue spheres represent lower hydraulic conductivity, which is often the case at greater depth. The dimensions are meters and the vertical exaggeration is 600:1. It is difficult to represent 3D data on a flat page, so you'll have to download the files for better visualization. TP2 has several features specifically designed to handle this type of data and it has already been processed. The results are stored in a unique file format in folder examples\geology. Drop the file dataset1.f3d onto TP2.exe (or launch it from a command line>TP2 dataset1f3d) to see the animation. You can move through the volume with the arrow keys. The animation will stop as soon as you press a key. A typical view is shown on the next page.

Geology.c is a 3D modification of solar.c plus ozone.c, which will create an animation that is a series of horizontal slices through the volume. The hydraulic conductivity "field" is approximated using Legendre polynomials. It could also be approximated using inverse distance or linearly interpolated after breaking the volume into tetrahedra. We first construct the approximation in 3D, then evaluate it one slice at a time to build the animation.

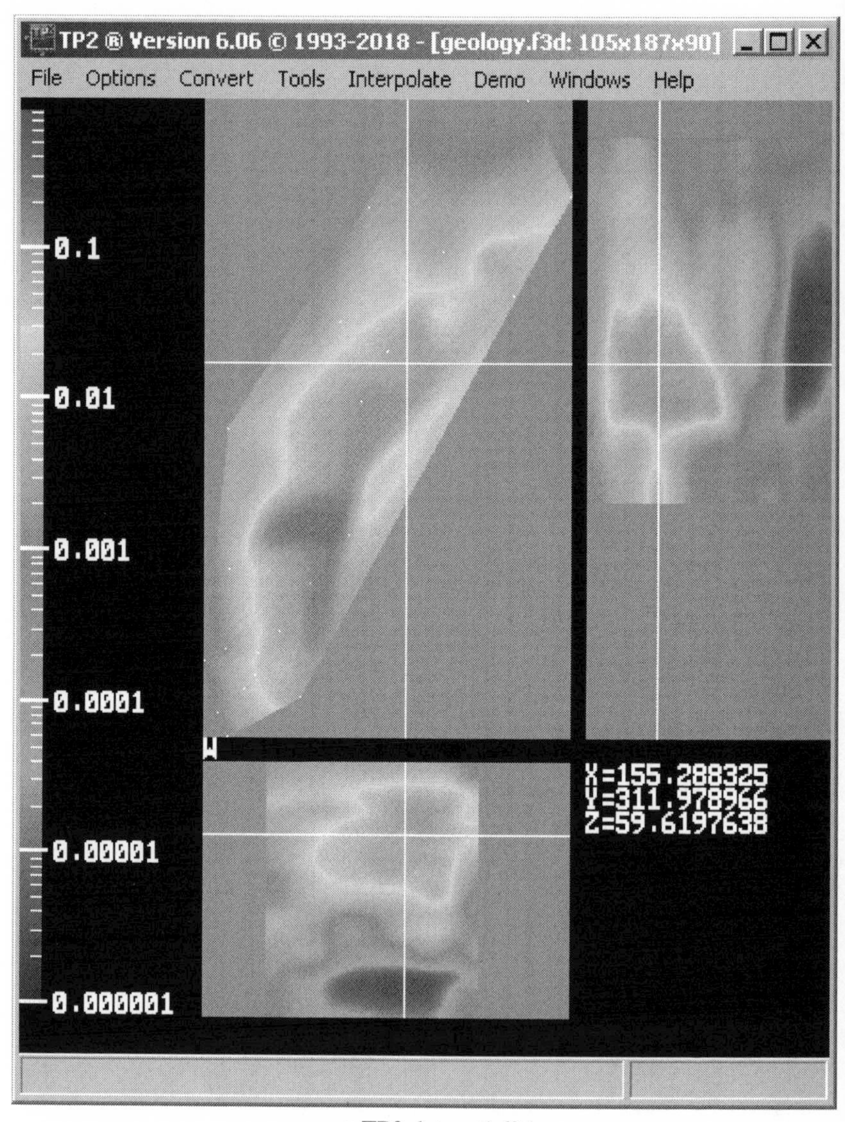

run>TP2 dataset1.f3d
or drop dataset1.f3d onto TP2.exe

Two of the more interesting horizontal slices through this field are shown in the following figures:

just below the surface

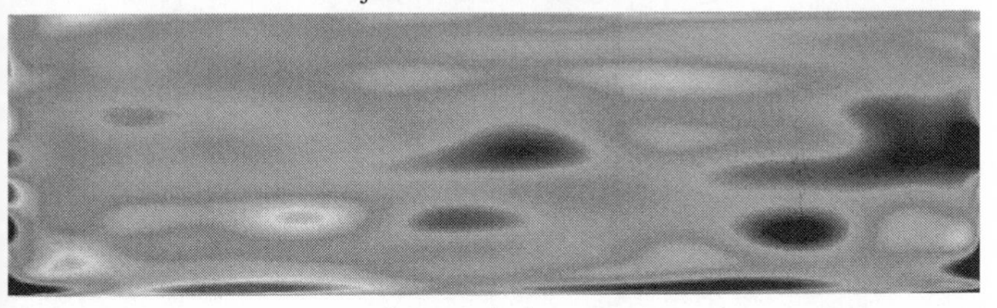

closer to the bottom

There are several data sets in the geology folder. Another way of displaying hydraulic conductivity data is shown in this next figure:

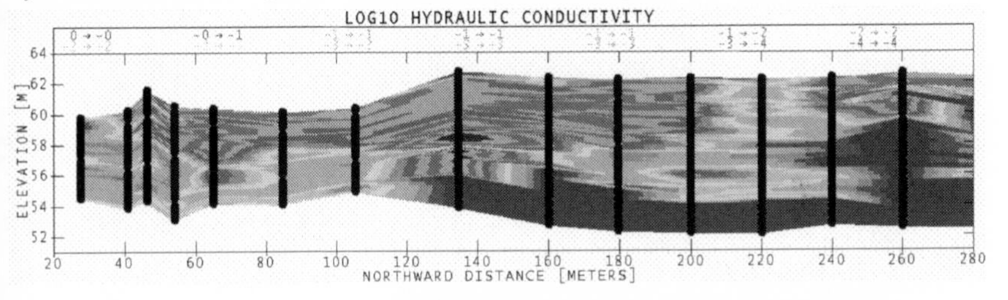

Limiting the Domain

Up until this point we have been rotating the data to orient it predominantly with either the X- or Y-axis, but this isn't always necessary. The data were not rotated or rescaled in the TP2 rendering shown previously. A bounding polygon is often used. If you don't know how to find a bounding polygon, see Appendix E for an illustration and code. The unrotated, unscaled, bounded code and data can be found in the same folder: examples\geology.

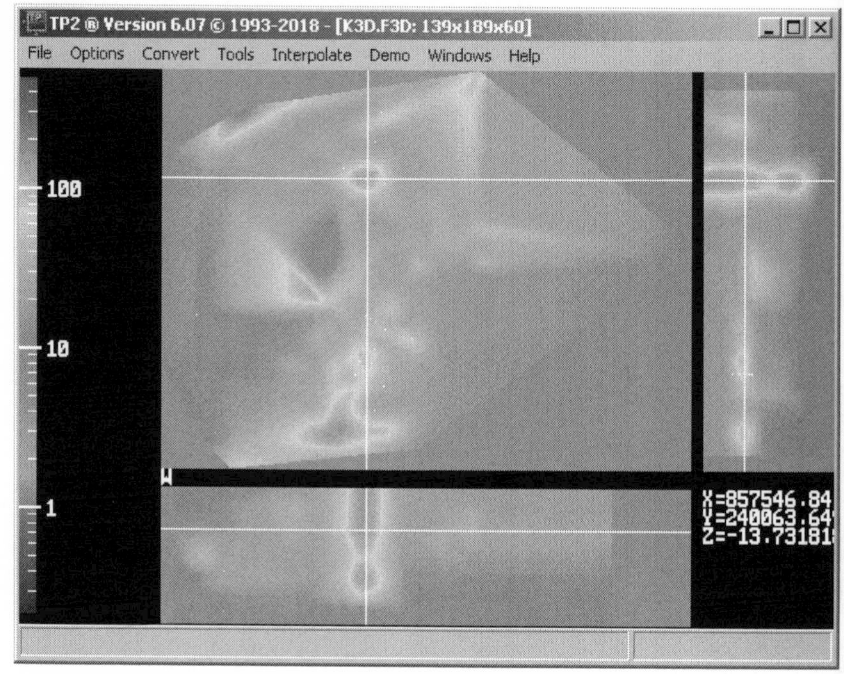

run>TP2 dataset2.f3d

or drop dataset2.f3d onto TP2.exe

The output of bounded.c is similar to geology.c, only the data representation is limited to the bounding polygon.

```
>bounded
reading data: dataset2.csv
  480 lines read
  479 data points found
  835937<X<872861
  210576<Y<250985
  -159.3<Z<54.2
  -0.66<K<2.69
creating approximation: 16x16x16
  4096 terms
bounding data
  polygon contains 12 points
animation: 548x600x16
rendering frame 1: Z=-159.3
rendering frame 2: Z=-145.067
rendering frame 3: Z=-130.833
rendering frame 4: Z=-116.6
rendering frame 5: Z=-102.367
rendering frame 6: Z=-88.1333
```

```
rendering frame 7: Z=-73.9
rendering frame 8: Z=-59.6667
rendering frame 9: Z=-45.4333
rendering frame 10: Z=-31.2
rendering frame 11: Z=-16.9667
rendering frame 12: Z=-2.73333
rendering frame 13: Z=11.5
rendering frame 14: Z=25.7333
rendering frame 15: Z=39.9667
rendering frame 16: Z=54.2
saving animation: bounded.gif
  size: 548x600x16
```

The first slice is shown below:

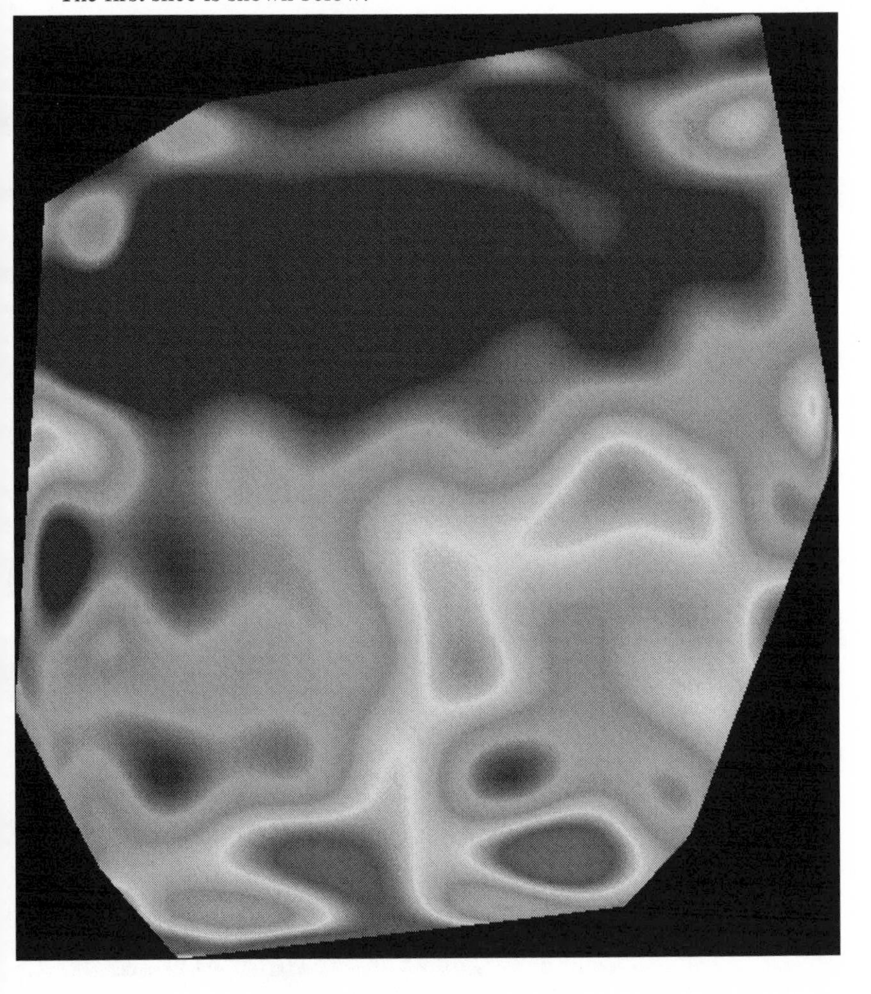

3D Plume Data

For three-dimensional contaminant data we return to using Hermite polynomials, as the concentrations obviously vanish at some distance away from the source. For this task we will modify the second geology code, bounded.c, with the limited domain (bounding convex polygon) and will again create slices to visualize the results. The files can be found in folder examples\Hermite\3D. The program is contained in plume3D.c and there are several data sets in this same folder. Consider the following concentration data taken from wells:

```
X,Y,Z,log10(C)
863769,238751,16,-1.177
863769,238751,19,-1.062
863769,238751,22,-0.941
863769,238751,25,-0.908
etc.
(see WAF.csv)
```

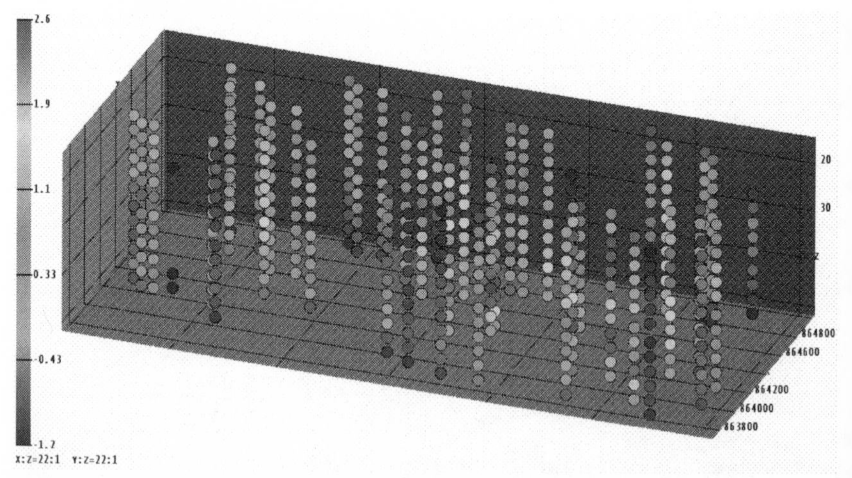

One slice through the field looks like:

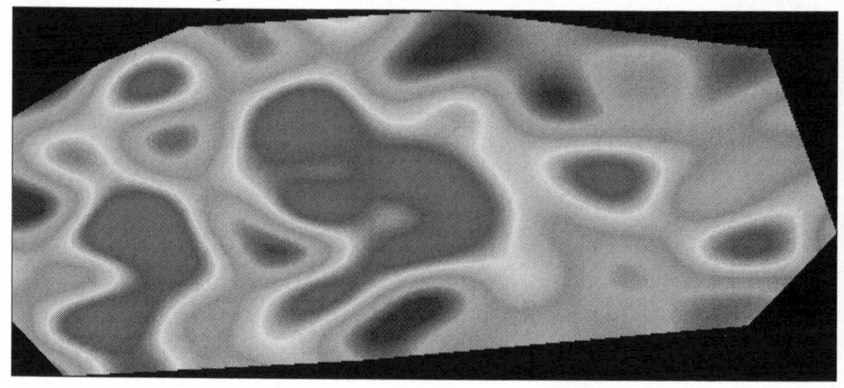

The sliced volume looks like:

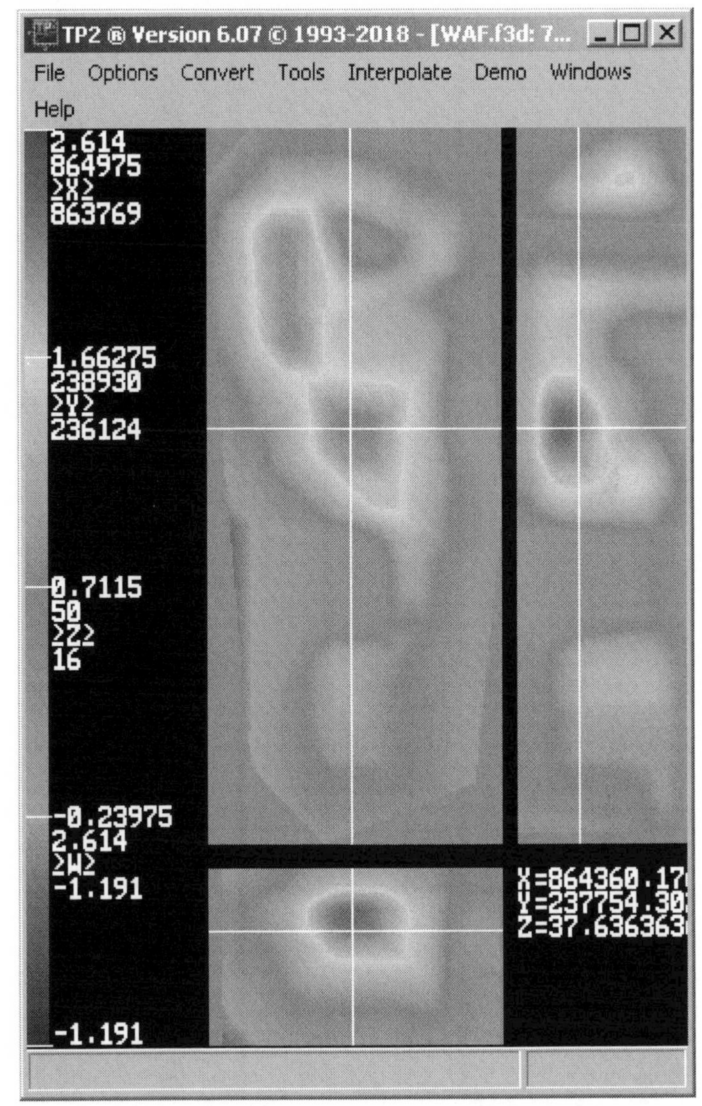

The color scale in plume3D is somewhat stretched beyond that in the figure above. You may want to eliminate (comment-out) the last 32 shades of magenta before the final RGB, which is white and should be retained. You can also stretch the min and max K to expand the coloration scale.

Chapter 8: Spherical Data

Data on the surface of a sphere must repeat itself. Values along the equator at +180° must be the same as -180° and also at 540° and -540°. If you start at the equator headed north and keep going past the North Pole, you will eventually reach the equator at the other side, then the South Pole, and finally back where you started after 360°. Therefore, sin() and cos() are orthogonal functions ideally suited to spherical data.

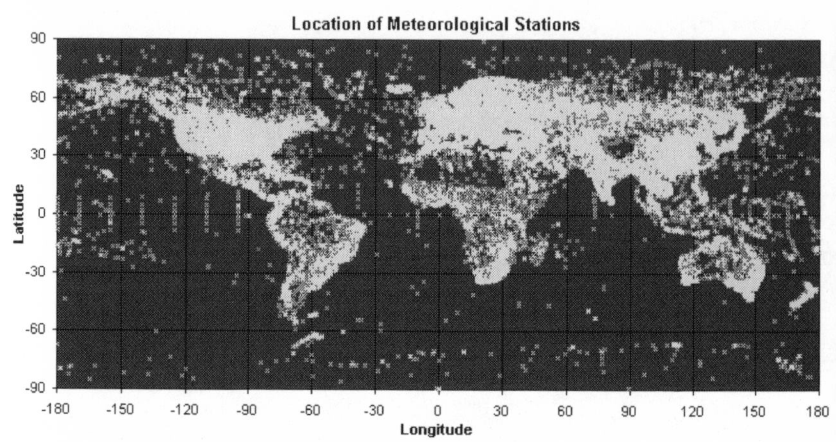

Earth data of many types is readily available. We will begin with atmospheric temperatures at the surface. There are approximately 28,000 meteorological stations that regularly report data, although only about 12,000 are active on any given day. These are shown in the preceding figure. These are mostly on land, but there are a few buoys. The data repository is maintained by the National Oceanographic and Atmospheric Administration (NOAA) on the National Climate Data Center (NCDC) ftp server in Asheville, North Carolina. You will find the data in the public area under Global Surface Summary of the Day (GSOD). Here's a link to the site:

ftp://ftp.ncdc.noaa.gov/pub/data/gsod/

The data are stored by station in compressed files that are combined into a tarball (i.e., a UNIX zip file) for each year. The quantity of data is daunting, but well worth the effort to dig in. I've been analyzing this climate data for over 25 years and have various tools to conveniently extract and organize the data. I have already prepared several data files, which you will find along with code in the folder examples\global. The program tglobal.c contains a variety of options for analyzing and displaying the data. You can change the switches, comment out function calls, and recompile to activate the many combinations.

Data Distribution

Up until this time, we've presumed the data are fairly evenly distributed so that a simple summation is sufficient to approximate an integral over the orthogonal region. This is definitely not the case with global temperature stations that are clustered near population centers and on land. This creates a significant problem for our approximation. The following figure shows the stations reporting on 7/1/2018 in a Mercator projection.

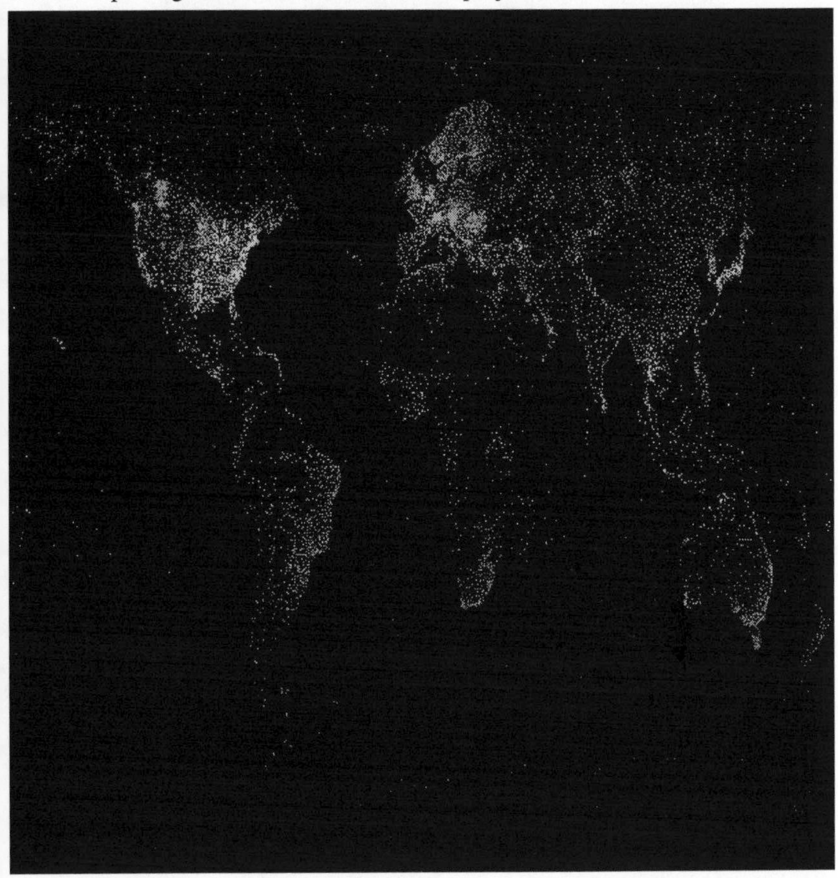

As expected, the specks near the equator are yellow, orange, and red, while those nearer the poles are blue.

Here's the same data shown in a Mollweide projection. The program will produce either one. To select, #define use_Mercator (or not).

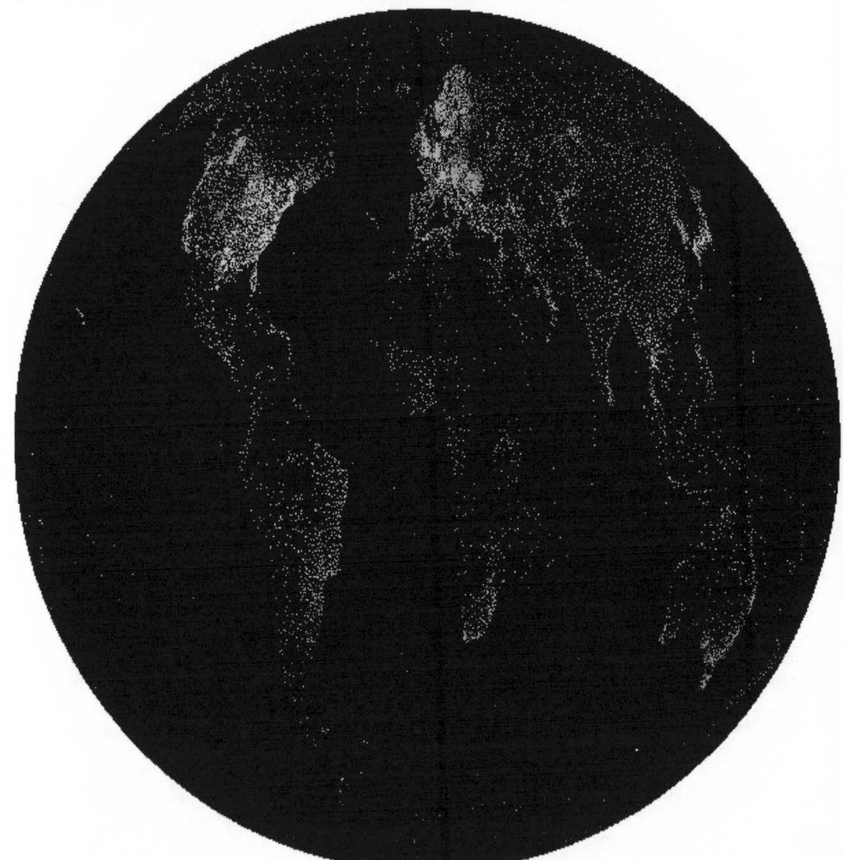

We can spread the sparse data near the poles outward (there is in reality only a single point at ±90°) and then fill in the gaps by averaging neighboring points that have data. In this way we can produce a uniformly-distributed grid from which to build an approximation. If we simply sweep through the grid, filling in neighboring points as they are encountered using the following simplistic algorithm:

```
do{ /* spread known temperatures outward */
  for(k=y=0;y<ny;y++)
    {
    for(x=0;x<nx;x++)
      {
      if(grid.n[nx*y+x])
        continue;
      for(T=n=0,i=y-1;i<=y+1;i++)
```

```
      {
      if(i<0)
        continue;
      if(i>=ny)
        break;
      for(j=x-1;j<=x+1;j++)
        {
        if(j<0)
          continue;
        if(j>=nx)
          break;
        if(grid.n[nx*i+j]==0)
          continue;
        n++;
        T+=grid.T[nx*i+j];
        }
      }
    if(n==0)
      continue;
    grid.n[nx*y+x]=1;
    grid.T[nx*y+x]=T/n;
    k++;
    }
  }
}while(k);
```

This leads to highly undesirable diagonal bleeding, as shown in this next figure:

This is a very unrealistic representation of the sparse data. The actual grid looks like the previous figure. I've spread them around in the figure above to make the bleeding clearer. A much more effective algorithm is:

```
for(y=0;y<ny;y++)
  {
  for(x=0;x<nx;x++)
    {
    if(grid.n[nx*y+x]>0)
      continue;
    for(r=k=1;k<5;r++)
      {
      T=k=0;
      for(i=y-r;i<=y+r;i++)
        {
        if(i<0)
```

```
          continue;
        if(i>=ny)
          break;
        for(j=x-r;j<=x+r;j++)
          {
          if(j<0)
            continue;
          if(j>=nx)
            break;
          if(grid.n[nx*i+j]<=0)
            continue;
          if((i-y)*(i-y)+(j-x)*(j-x)>r*r)
            continue;
          k++;
          T+=grid.T[nx*i+j]/abs(grid.n[nx*i+j]);
          }
        }
      }
    grid.n[nx*y+x]=-k;
    }
  printf("\r%.01f%%",(y+1)*100./ny);
  }
```

You can select between the two with #define bleed (or not).

Instead, we must work radially outward from each point, which produces the pattern shown in this next figure:

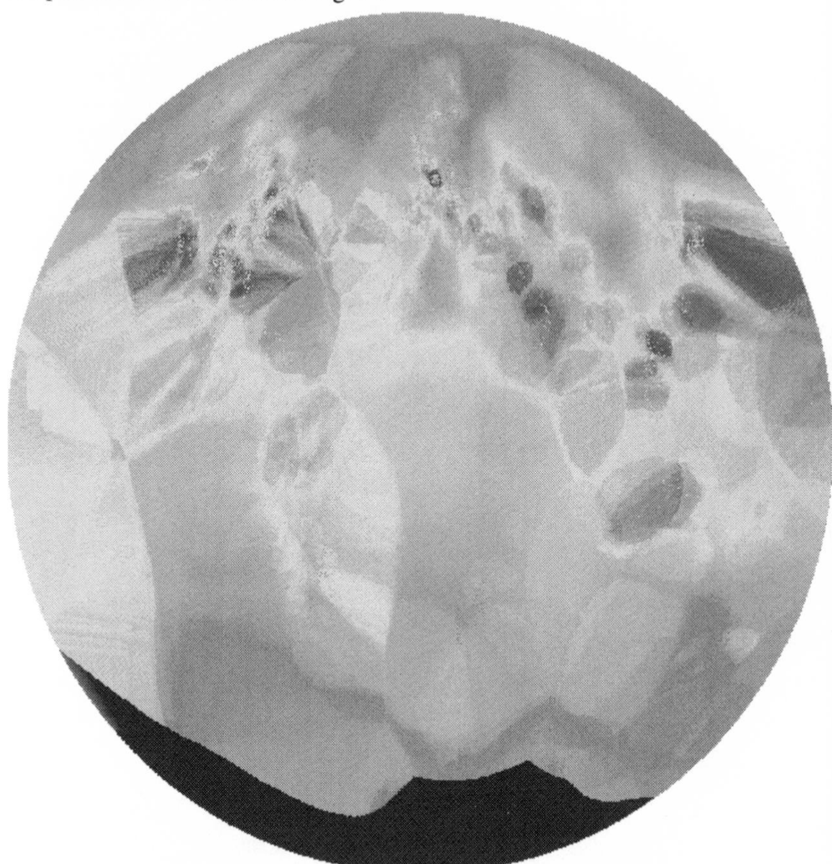

This takes much longer, but is a much more realistic representation of the sparse data. Needless to say, this uniformly-distributed is much more easily approximated. Recall the orthogonality interval for sin() and cos() is $-\pi$ to $+\pi$, so you must multiply the latitude by 2. The temperature data are read from the file specified in argv[1], so that you can drop this on tglobal.exe or launch it with a command line argument. The program outputs one or more GIFs.

```
 ReadStations("stations.csv");
 ReadTemperatures(argv[1]);
 DistributeTemperatures(90,90);
#ifdef use_Fourier
 FourierApproximation(12,12);
```

```
#else
 LeastSquares(9);
#endif
#ifdef use_Mercator
 MercatorProjection(600,600,1,0,"Mercator1.gif");
 if(grid.T)
  MercatorProjection(600,600,0,-1,"Mercator2.gif");
 else
  MercatorProjection(600,600,-1,0,"Mercator2.gif");
#else
 MollweideProjection(600,600,1,0,"Mollweide1.gif");
 if(grid.T)
  MollweideProjection(600,600,0,-1,"Mollweide2.gif");
 else
  MollweideProjection(600,600,-1,0,"Mollweide2.gif");
#endif
```

This next figure shows the 12x12 (144-term) Fourier approximation for 7/1/2018. There are a few artifacts (reflections, blotches, shadows, etc.), but an overall acceptable result.

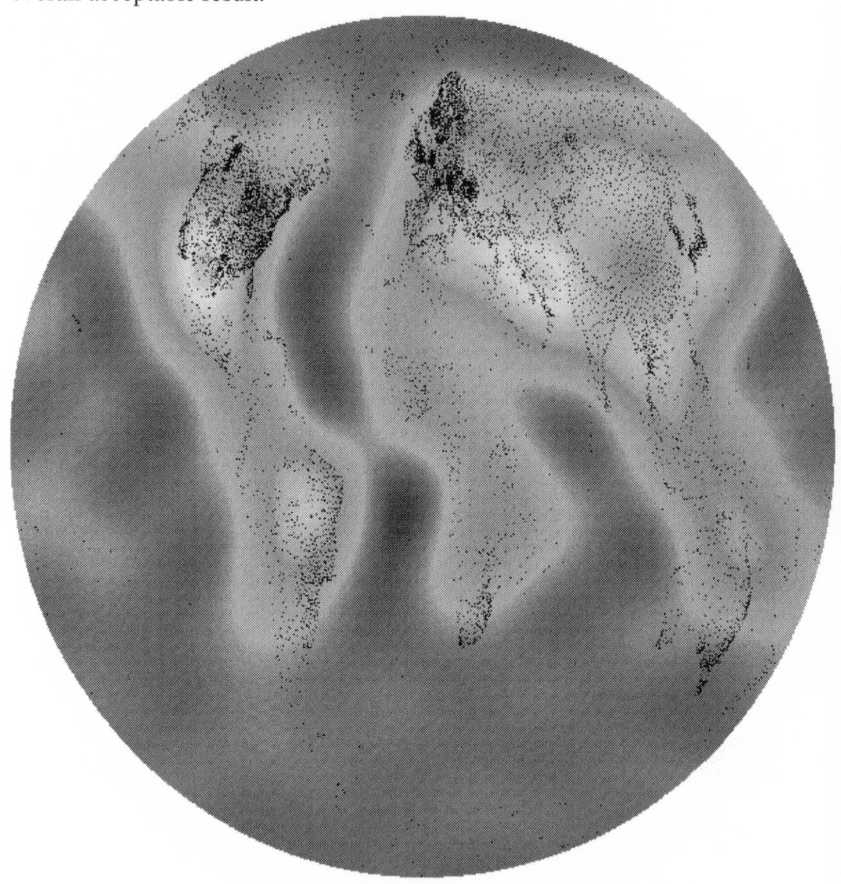

The option of using lat, lon, lat^2, lat*lon, lon^2, etc. is also included and controlled by #define use_Fourier (or not). This next figure is the 9-th order (55-term) binomial approximation for the same day. The binomial approximation contains far more artifacts. Lower orders have fewer artifacts but are clearly inadequate to represent the variability. Higher orders have increasingly more unwanted artifacts, which is why the binomial approximation is unacceptable.

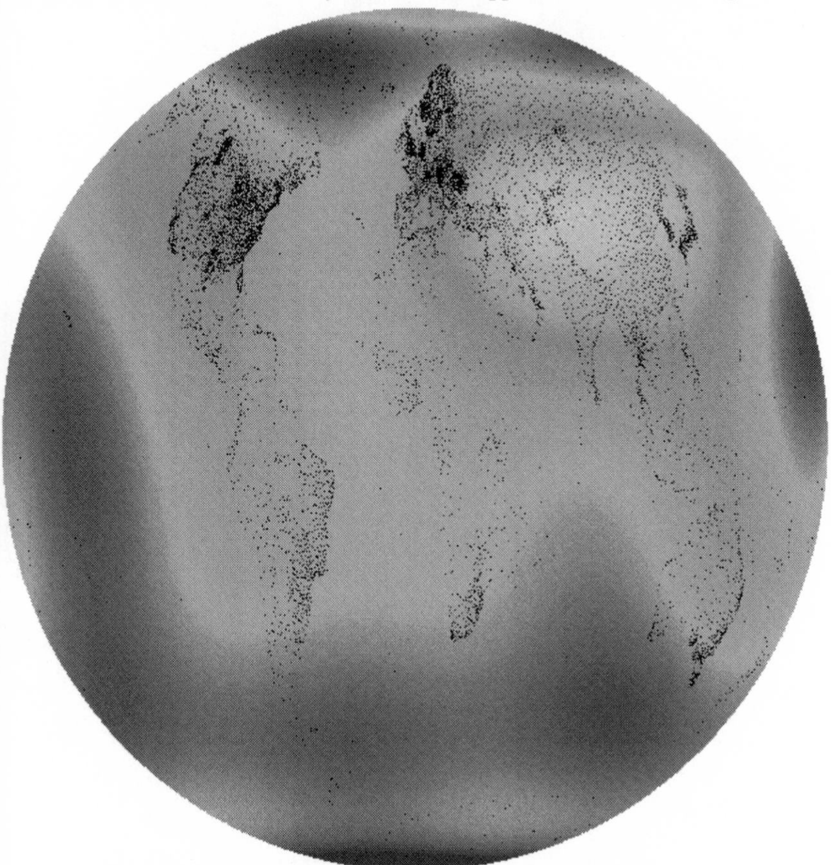

The daily files also contain dew-point (also in ºF), barometric pressure (in mbar), rain (inches), and wind (knots).[16] You can modify the temperature map code to map these other variables. The dry-bulb is most extensively reported. The other variables not so much. Bad or missing values are indicated by 999s.

---

[16] The NCDC chose these units.

The humidity can be calculated from the dry-bulb and dew-point to arrive at the following figure:

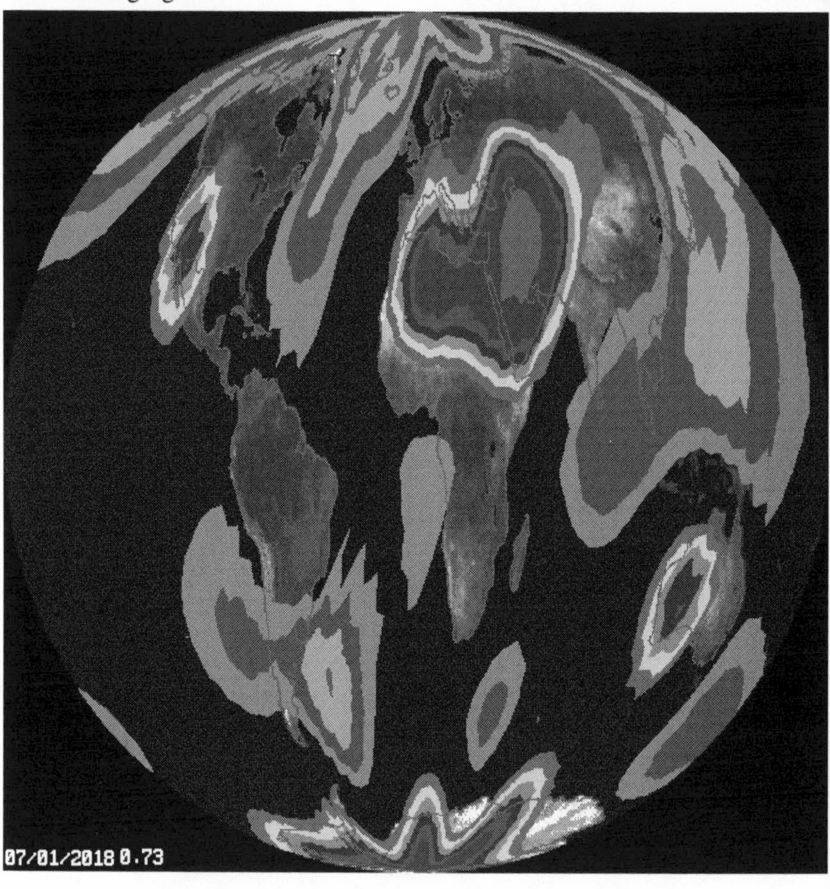

The rainfall shown in this figure coincides with a huricane in the South Pacific.

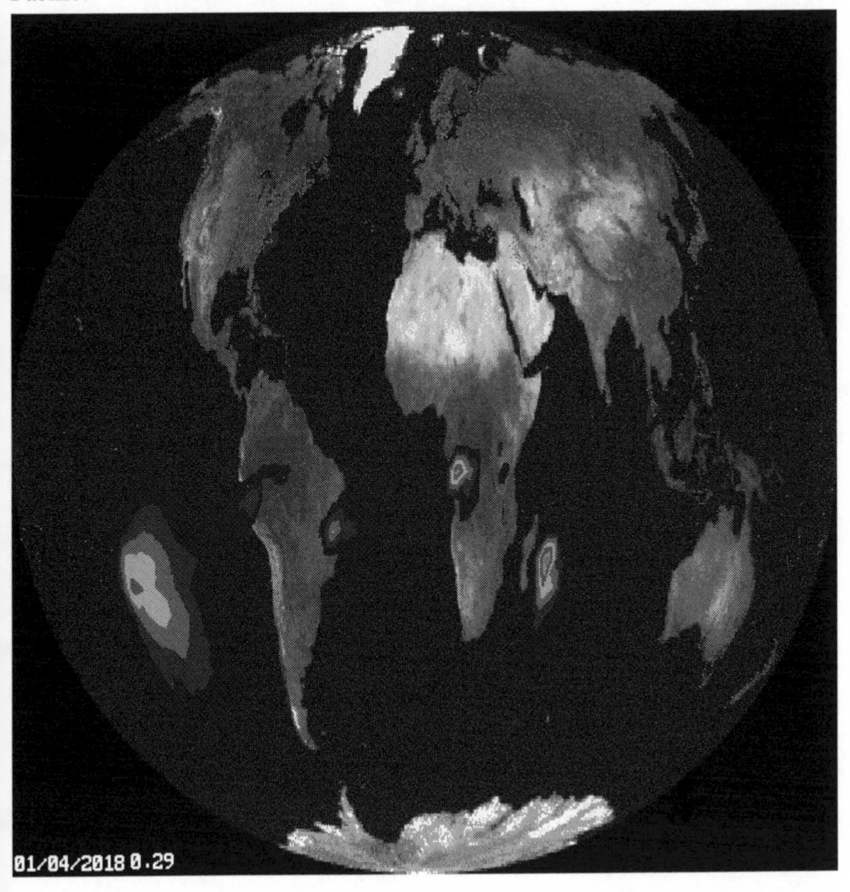

Contours of barometric pressure are shown in this next figure:

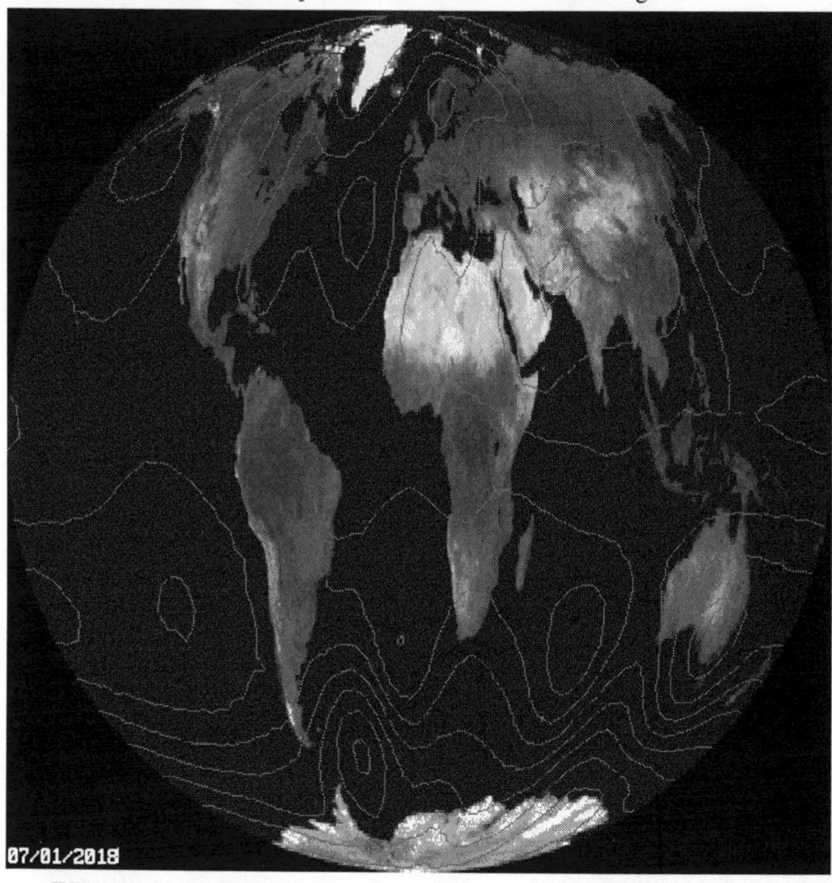

Wind is driven by pressure. Wind blows in the direction of the negative gradient of the pressure. The wind vectors are a combination of the wind data plus the gradient of the barometric pressure data. The following code draws the barometric pressure contours and also calculates the wind vectors:

```
void DrawCont(GRID*g1,GRID*g2,GRID*g3,GRID*g4,double
   P,BYTE color)
  {
  int i1,i2,j1,j2;
  if((P-g1->P)*(P-g2->P)>=0.)
    return;
  if((P-g3->P)*(P-g4->P)>=0.)
    return;
  if(g2->i==g1->i)
    i1=g1->i;
  else if(fabs(g1->P-g2->P)<FLT_EPSILON)
```

```
    i1=(g1->i+g2->i)/2;
  else
    i1=nint(g1->i+(g2->i-g1->i)*(P-g1->P)/(g2->P-g1-
   >P));
  if(g2->j==g1->j)
    j1=g1->j;
  else if(fabs(g1->P-g2->P)<FLT_EPSILON)
    j1=(g1->j+g2->j)/2;
  else
    j1=nint(g1->j+(g2->j-g1->j)*(P-g1->P)/(g2->P-g1-
   >P));
  if(g4->i==g3->i)
    i2=g3->i;
  else if(fabs(g3->P-g4->P)<FLT_EPSILON)
    i2=(g3->i+g4->i)/2;
  else
    i2=nint(g3->i+(g4->i-g3->i)*(P-g3->P)/(g4->P-g3-
   >P));
  if(g4->j==g3->j)
    j2=g3->j;
  else if(fabs(g3->P-g4->P)<FLT_EPSILON)
    j2=(g3->j+g4->j)/2;
  else
    j2=nint(g3->j+(g4->j-g3->j)*(P-g3->P)/(g4->P-g3-
   >P));
  DrawLine(i1,j1,i2,j2,color);
  }

void DrawContour(GRID*g1,GRID*g2,GRID*g3,double P,BYTE
   color)
  {
  DrawCont(g1,g2,g2,g3,P,color);
  DrawCont(g1,g2,g1,g3,P,color);
  DrawCont(g1,g3,g2,g3,P,color);
  }

void DrawContours(GRID*g1,GRID*g2,GRID*g3,double
   Pm,double Px,double dP,BYTE color)
  {
  double P;
  for(P=Pm;P<=Px;P+=dP)
    DrawContour(g1,g2,g3,P,color);
  }

void DrawVectors(GRID*g1,GRID*g2,GRID*g3,BYTE color)
  {
  int i,i1,i2,i3,i4,j,j1,j2,j3,j4,k,l;
  double angle,D11,D12,D13,D21,D22,D23,D31,D32,D33,
   Det,lat,Po,Px,Py,V,W;
```

```
W=(g1->W+g2->W+g3->W)/3.;
if(W<0.)
  return;
V=W/3.;
if(V>6.)
  V=6.;//

Det=(g2->lon-g1->lon)*(g3->lat-g2->lat)-
    (g3->lon-g2->lon)*(g2->lat-g1->lat);
if(fabs(Det)<FLT_EPSILON)
  return;

D11=(g2->lon*g3->lat-g3->lon*g2->lat)/Det;
D12=(g3->lon*g1->lat-g1->lon*g3->lat)/Det;
D13=(g1->lon*g2->lat-g2->lon*g1->lat)/Det;
D21=(g2->lat-g3->lat)/Det;
D22=(g3->lat-g1->lat)/Det;
D23=(g1->lat-g2->lat)/Det;
D31=(g3->lon-g2->lon)/Det;
D32=(g1->lon-g3->lon)/Det;
D33=(g2->lon-g1->lon)/Det;
Po=g1->P*D11+g2->P*D12+g3->P*D13;
Py=g1->P*D21+g2->P*D22+g3->P*D23;
Px=g1->P*D31+g2->P*D32+g3->P*D33;

if(fabs(Px)+fabs(Py)<FLT_EPSILON)
  return;

lat=(g1->lat+g2->lat+g3->lat)/3.;
Px*=cos(radians(lat));
V*=cos(radians(lat));

angle=atan2(-Py,-Px);
j=(g1->j+g2->j+g3->j)/3;
i=(g1->i+g2->i+g3->i)/3;

k=nint(cos(angle)*V);
l=nint(sin(angle)*V);
j1=j-k;
i1=i-l;
j2=j+k;
i2=i+l;
k=nint(cos(angle+150.*M_PI/180.)*3.);
l=nint(sin(angle+150.*M_PI/180.)*3.);
j3=j2+k;
i3=i2+l;
k=nint(cos(angle-150.*M_PI/180.)*3.);
l=nint(sin(angle-150.*M_PI/180.)*3.);
```

```
  j4=j2+k;
  i4=i2+l;

  DrawLine(i1,j1,i2,j2,color);
  DrawLine(i2,j2,i3,j3,color);
  DrawLine(i2,j2,i4,j4,color);
  }

void CreateWindMap(int year,int month,int day)
  {
  char fname[13],text[13];
  int aday,d,done,i,j,k,n,s;
  double co,P,W,Wa,Wc,time1,time2;
  FILE*fo;
  SYSTEMTIME begin,end;

  GetSystemTime(&begin);
  time1=begin.wMilliseconds/1000.+begin.wSecond
       +60.*(begin.wMinute+60.*begin.wHour);

/* initialize image */

  memcpy(Image,BMPimage(Earth),image_high*image_wide);

/* build list of stations having data for this day */

  aday=Julian(year,month,day)-Julian(year,1,1);
  for(n=s=0;s<stations;s++)
    if(Station[s].pres[aday]<9999.)
      has_data[n++]=s;
  has_data[n]=-1;
  if(n<3)
    return;

/* interpolate pressures and wind */

  for(Wa=Wc=0.,done=i=0;i<lats;i++)
    {
    for(j=0;j<lons;j++)
      {
      P=InterpolatePressure(Grid[i][j].lat,
        Grid[i][j].lon,aday);
      Grid[i][j].P=P;
      W=InterpolateWind(Grid[i][j].lat,
        Grid[i][j].lon,aday);
      Grid[i][j].W=W;
      co=cos(radians(Grid[i][j].lat));
      Wa+=co*W;
      Wc+=co;
```

```
      }
    d=nint(i*100./lats);
    if(d>done)
      printf("%i%% done\r",done=d);
    }
  Wa/=Wc;

/* relax (smooth) pressure field */

  for(k=0;k<smooth;k++)
    {
    for(i=0;i<lats;i++) /* copy current values to
   temporary variables */
      for(j=0;j<lons;j++)
        Grid[i][j].Q=Grid[i][j].P;
    for(P=j=0;j<lons;j++) /* south pole */
      P+=Grid[0][j].Q;
    P/=lons;
    for(j=0;j<lons;j++)
      Grid[0][j].P=P;
    for(i=1;i<lats-1;i++) /* mid latitudes */
      {
      for(j=0;j<lons;j++)
        {
        P=Grid[i-1][j].Q+Grid[i+1][j].Q;
        if(j>0)
          P+=Grid[i][j-1].Q;
        else
          P+=Grid[i][lons-1].Q;
        if(j<lons-1)
          P+=Grid[i][j+1].Q;
        else
          P+=Grid[i][0].Q;
        Grid[i][j].P=P/4.;
        }
      }
    for(P=j=0;j<lons;j++) /* north pole */
      P+=Grid[lats-1][j].Q;
    P/=lons;
    for(j=0;j<lons;j++)
      Grid[lats-1][j].P=P;
    }

/* identify stations */

  if(show_stations!=-1)
    {
    for(k=0;has_data[k]>=0;k++)
      {
```

```
      s=has_data[k];
      i=Station[s].i;
      j=Station[s].j;
      Image[image_wide*i+j]=show_stations;
      }
    }

/* draw continental polygons */

  if(show_continents!=-1)
    for(k=0;k<polys;k++)
      for(i=Poly[k].n-1,j=0;j<Poly[k].n;i=j++)
        DrawLine(Poly[k].i[i],Poly[k].j[i],
          Poly[k].i[j],Poly[k].j[j],show_continents);

/* draw grid */

  if(show_grid!=-1)
    {
    for(i=0;i<lats-1;i++)
      {
      for(j=0;j<lons-1;j++)
        {
        DrawLine(Grid[i  ][j  ].i,Grid[i  ][j  ].j,
                 Grid[i+1][j  ].i,
                 Grid[i+1][j  ].j,show_grid);
        DrawLine(Grid[i  ][j  ].i,Grid[i  ][j  ].j,
                 Grid[i  ][j+1].i,
                 Grid[i  ][j+1].j,show_grid);
        }
      }
    }

/* draw pressure contours */

  if(show_contours!=-1)
    {
    for(i=0;i<lats-1;i++)
      {
      for(j=0;j<lons-1;j++)
        {
        DrawContours(&Grid[i][j],&Grid[i+1][j  ],
          &Grid[i+1][j+1],900.,1150.,5.,show_contours);
        DrawContours(&Grid[i][j],&Grid[i+1][j+1],
          &Grid[i  ][j+1],900.,1150.,5.,show_contours);
        }
      }
    }
```

```
/* draw wind vectors */

  if(show_vectors!=-1)
    {
    for(i=0;i<lats-1;i++)
      {
      for(j=0;j<lons-1;j++)
        {
        DrawVectors(&Grid[i][j],&Grid[i+1][j  ],
          &Grid[i+1][j+1],show_vectors);
        DrawVectors(&Grid[i][j],&Grid[i+1][j+1],
          &Grid[i  ][j+1],show_vectors);
        }
      }
    }

/* embed date and average wind velocity */

  sprintf(text,"%02i/%02i/%04i",month,day,year);
  EmbedText(2,14,text,white,black,1);

  sprintf(text,"%4.1lf",Wa);
  if(image_wide==360)
    EmbedText(318,14,text,white,black,1);
  else
    EmbedText(678,14,text,white,black,1);

/* save image */

  sprintf(fname,"%04i%02i%02i.bmp",year,month,day);
  printf("writing image %s",fname);
  BMPwrite(Globe,fname);

/* append average pressure to the list */

  if((fo=fopen("windmap.csv","at"))==NULL)
    Abort(__LINE__,"can't update output file");
  fprintf(fo,"%i/%i/%i,%lf,%i\n",month,day,year,Wa,n);
  fclose(fo);

  GetSystemTime(&end);
  time2=end.wMilliseconds/1000.+end.wSecond
       +60.*(end.wMinute+60.*end.wHour);
  if(time2<time1)
    time2+=3600.;
  printf(" Wavg=%4.1lf, n=%i, sec=%.0lf\n",Wa,n,time2-
   time1);
  }
```

Wind vectors are shown in this next figure:

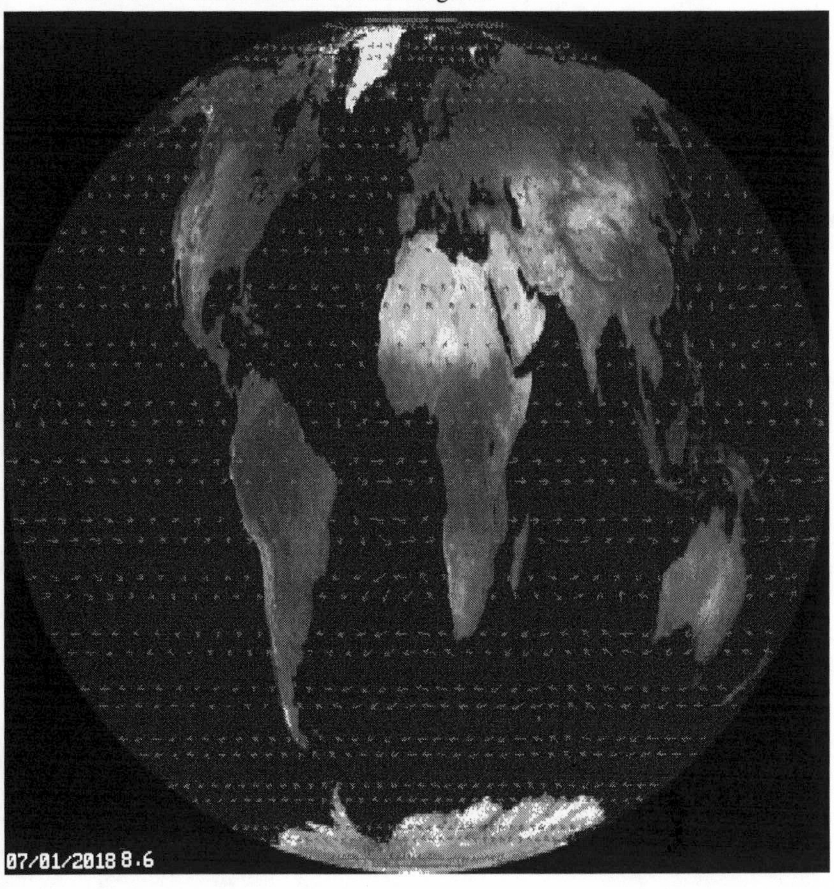

Appendix A: Discrete Radon Transform

There are many discussions on the Web of the Radon Transform. So many, in fact, that it can be quite confusing. Some mention Fast Fourier Transform (FFT) and some do not, while others mention it and then never use it. In regards to the Radon Transform, there are two streams of information that started at the same point but are now completely separate. This transform was first introduced by Radon in 1917.[17] It has been employed in several areas, ranging from medical scanners to seismic event reconstruction. The original motivation for this transform and initial applications dealt with processing an analog signal in order to render a two-dimensional image. Three-dimensional applications were added later. This transform is inexact and has no exact inverse, that is, there is no way to render the image precisely.

Some time later, researchers pursuing other goals began using the Radon Transform—independent of the FFT. These researchers began with an image—something that isn't possible with a CAT scan or seismograph. The most complete work on this subject has been done by William H. Press. His paper, "Discrete Radon Transform Has an Exact, Fast Inverse and Generalizes to Operations other than Sums along Lines," is readily available on the Web and well worth reading. While the Discrete Radon Transform (DRT) described by Press uses the same modified polar coordinates of Radon[18], what follows after that has nothing to do with the former.

Of course, when you start with an image and apply a simple polar transform on a pixel-by-pixel level there will be an exact inverse—so long as information wasn't lost in the transformation. This has nothing to do with digitizing and analyzing an analog signal to arrive at a two-dimensional image. Of course, FFTs and their inverse are not required because there's no signal. The source is already an image and the forward transform is the same pixels moved to different locations. While there may be some interesting applications, such a filtering in the transform domain, the DRT doesn't fit into our discussion of orthogonal functions.

There is also a lot of code available on the Web to perform a DRT. Most of this code, however, is useless, as it will only compile on Linux with the GNU system or will only run in Matlab or requires some FFT library that is incompatible with 98.5% of the computers currently on the Earth. I have provided a small—and completely self-contained—program that performs this

---

[17] Radon, J., "Über die Bestimmung von Funktionen durch ihre Integralwerte läangs gewisser Mannigfaltigkeiten," Sächsische Akademie der Wissenschaften ["On the Determination of Functions by Their Integral Values along the Course of Certain Manifolds," Saxon Academy of Sciences], Leipzig 69, pp. 262-277, 1917.

[18] The conventional polar transform is based on radius, r, and angle theta, θ, and spans $r=0$ to ∞ and $\theta=0$ to 2π. The Radon Transform uses $r=-1$ to $+1$ and $\theta=0$ *to* π. Otherwise, it's the same.

simple transformation. It easily compiles with the Microsoft® or Digital Mars® compilers and requires nothing else. It comes with two additional source code modules to read and write GIFs and JPEGs, respectively. These also readily compile and will run on any version of the Windows® operating system. The files and sample images can be found in the online archive in folder examples\DRT. The first image comes from Press' paper:

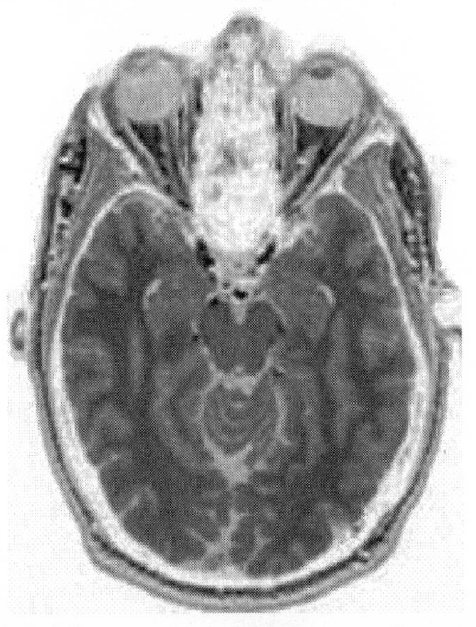

The transform image is:

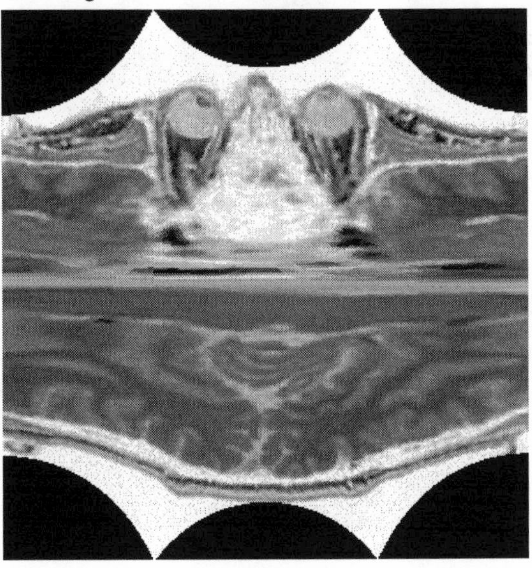

The inverse image is (almost exactly the same as the original):

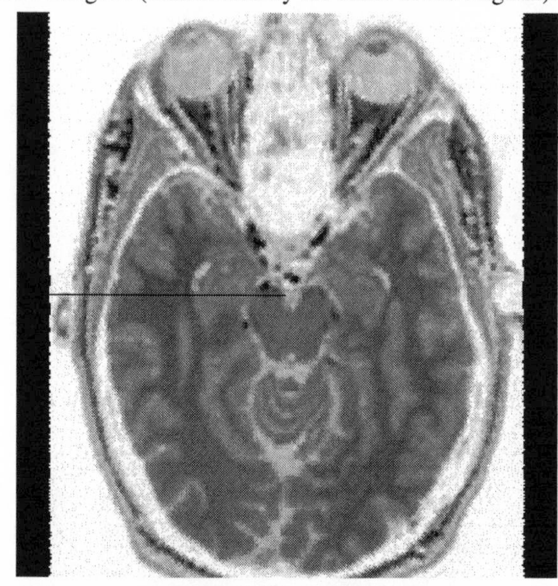

The second image used by Press is:

The transform of this image:

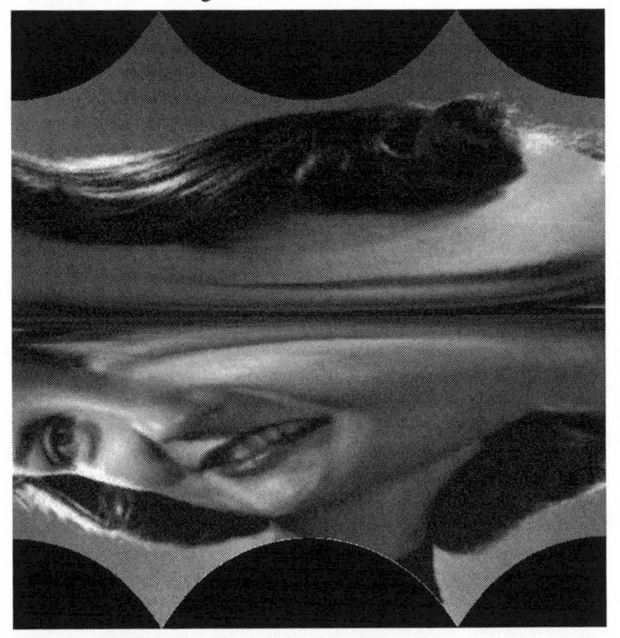

The inverse is nearly precise:

The code will read a BMP, GIF, or JPEG and output any of the same. The transform is quite simple:

```
for(h=0;h<bj->biHeight;h++)
  {
  r=(h-bj->biHeight/2)*hypot(bi->biHeight,bi-
 >biWidth)/bj->biHeight;
  for(w=0;w<bj->biWidth;w++)
    {
    t=w*M_PI/bj->biWidth;
    j=bi->biWidth/2+(int)(r*cos(t));
    i=bi->biHeight/2+(int)(r*sin(t));
    if(i>=0&&i<bi->biHeight&&j>=0&&j<bi->biWidth)
      jbits[jwide*h+w]=ibits[iwide*i+j];
    }
```

The inverse is also quite simple:

```
for(i=0;i<bj->biHeight;i++)
  {
  for(j=0;j<bj->biWidth;j++)
    {
    if(i-bi->biHeight/2!=0||j-bi->biWidth/2!=0)
      t=atan2(i-bi->biHeight/2,j-bi->biWidth/2);
    else
      t=0.;
    w=(int)(bj->biWidth*(fabs(t)/M_PI));
    r=hypot(j-bi->biWidth/2,i-bi->biHeight/2);
    h=(int)(bj->biHeight*(r/hypot(bi->biWidth,bi-
 >biHeight)+0.5));
    if(t<0.)
      {
      h=bi->biHeight-h-1;
      w=bi->biWidth-w-1;
      }
    if(h>=0&&h<bi->biHeight&&w>=0&&w<bi->biWidth)
      jbits[jwide*i+j]=ibits[iwide*h+w];
    }
  }
```

Appendix B. Image Rotation in Windows®

The proliferation of quirky, incompatible libraries has reached epidemic proportion. These require everything from Matlab® to Java®, are composed in such superfluous scripts as Python® and litter the Web. The ability to rotate images has been a native feature of the Windows® operating system from the beginning. You don't need an image library to perform this simple task. I have provided a compact application to illustrate this functionality. You will find it in the folder examples\rotate. The core function call is:

```
PlgBlt(rdc,q,cdc,0,0,bi.bmWidth,bi.bmHeight,NULL,0,0);
```

The rest of the code throws up a window, loads the image, and rotates it using a timer. This application also illustrates how to eliminate flicker. If you paint anything more complicated than an image having the same characteristics as the desktop, you will get flicker. To avoid this, first paint into a memory context and then perform a block transfer to the window using BitBlt(). Here's what it looks like:

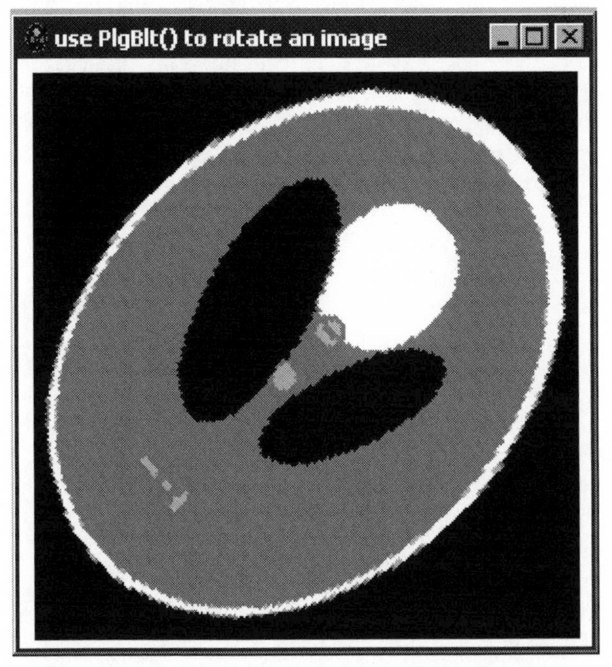

Windows® does not provide functions to import or export GIFs or JPEGs. These file formats include compression, can save a lot of space, and are quite useful. I have provided code do perform these operations in the folder examples\ utilities. The same are also used in some of the other examples.

Appendix C. Generation of Sinogram Images

Matlab® has built-in functions for the forward and reverse Radon Transform (radon and iradon). Clearly, these do not perform the literal pixel-by-pixel rearrangement of the original image as discussed by William Press and presented in Appendix A. Systems architect Amarnath S, Ph.D., has provided an excellent contribution to the CodeProject entitled, "Computed Tomography - Generation of the Sinogram In C#," which includes a bit of code plus 8 examples of before and after the transform is applied. Amarnath's comparison figures produced using Matlab's radon() function provide important missing information—not at all apparent from the Matlab® documentation.

Algorithms in pseudocode abound on the Web. The problem is, that many of these don't produce the result as described. There are a few challenges to reproducing the results shown in Amarnath's article, including: 1) accumulating the samples, 2) scaling the intensities, and 3) orienting the final images. First, if you accumulate the samples along a diagonal line (for example, using Bresenham's Algorithm, which would be equivalent to a horizontal traverse), rather than first rotating the image and summing vertically, you get completely different results. Second, grayscale bitmap images consist of BYTE codes ranging from 0x00 to 0xFF. These aren't floating-point numbers, they're offsets into a standard palette consisting of DWORDs from 0x000000 to 0xFFFFFF. Scaling of the final intensities is in no way automatic. Third, if you implement the loops provided in Amarnath's article, the images are mirrored (i.e., flipped left-to-right).

Code that actually produces the requisite images is:

```
for(j=0;j<180;j++) /* second pass save the image */
  {
  if(j)
    ro=BMProtate(bi,(double)j,black);
  else
    ro=bi;
  rbits=BMPbits(ro);
  rwide=BMPwidth(ro->biWidth,ro->biBitCount);
  for(i=0;i<si->biHeight;i++)
    {
    for(sum=h=0;h<ro->biHeight;h++)
      sum+=rbits[rwide*h+i];
    sbits[180*i+179-j]=(BYTE)((255*sum)/bigsum);
    }
  if(ro!=bi)
    free(ro);
  }
```

The same loop is used to get the largest sum (i.e., bigsum), which is then used to scale the intensities (i.e., (255*sum)/bigsum) so that the brightest pixel will be 0xFF. The index [180*i+179-j] flips the final image left-to-right.

Bitmaps are not some vague collection of numbers thrown together any way you like that automatically adjust themselves for optimum viewing. Bitmaps are defined by the Windows® API, have very specific content, and clearly defined ordering. Pixel [0,0] is the upper-left corner, not the lower left. They are ordered top-to-bottom and left-to-right. You will find the code (sinogram.c) and all of the images in folder examples\sinogram. Two comparisons follow:

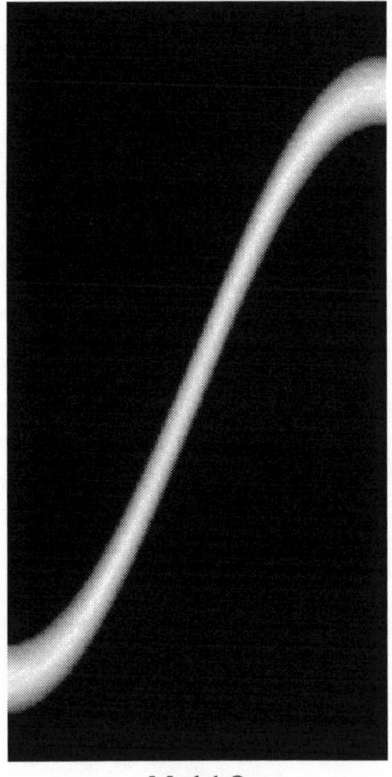

Matlab®

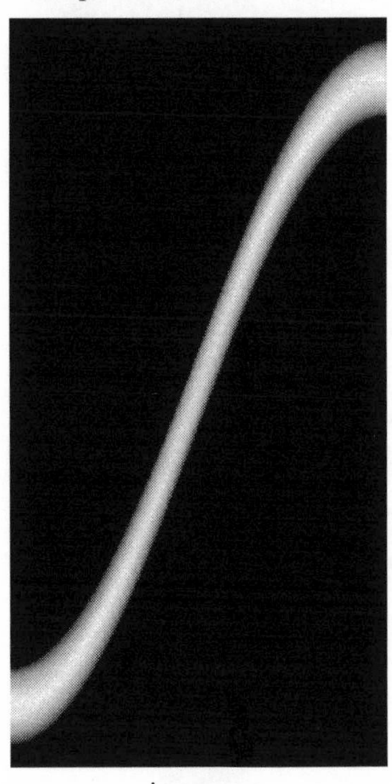

sinogram.c

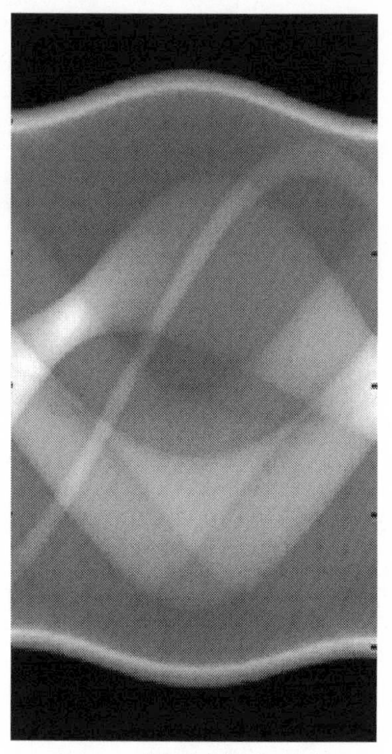

Matlab®

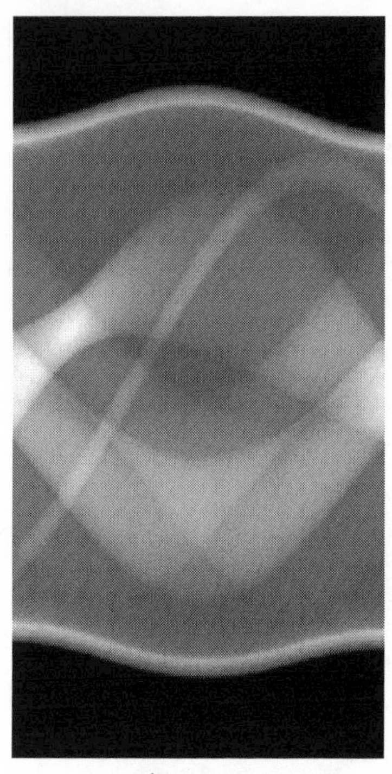

sinogram.c

Appendix D. Play It Anyway!

When CD drives first became available for PCs, I was working as part of a team on a large project for the Department of Defense. Some of these early CD drives had what was an oversight, rather than an intentional feature. These special drives would play whatever you put in them, that is, sound would come out the aux jack even while the drive was reading a data file. Most of the time this was a very annoying hissing or buzzing noise—but not always. A teammate, who listened to music while working, discovered this feature quite by accident.

The work was split up between several contractors around the United States. The latest official project data files (drawings, lists, chemical analyses, etc.) were sent out periodically to all of the contractors. This very special data CD actually produced an eerie modernistic sort of music! This weird noise immediately became the official theme song of the project. Word of this anomaly quickly spread across the country and through the halls of the Pentagon. Not everyone—just those with a sense of humor—from lawyers to ranking officers involved with the project wanted to hear the theme song.

They all had access to the CD (which was classified information), but not necessarily a device that would play it. We never found a car system that would play the data discs. We also never found another CD that produced anything quite like this special one. We did find a few drives that would play the disc too fast or too slow or the wrong option of stereo/mono, which completely spoiled the effect.

After much deliberation (actually, it was just a smirk) and a very lengthy planning session (about two seconds), I proposed a solution to address this pressing and most urgent issue of national security: a program to play anything, whether it was meant to be played or not! Thus was born, PlayIt.c, which you will find in the folder examples\playit. PlayIt is a small but complete Windows® program. It will play whatever phone number you type in using the IFT, as described in Chapter 2, or save it as a .wav file. You can open any file and play it, as well as select sampling rates and mono or stereo modes. You can also drop a file onto the .exe and it will play it. In this way, PlayIt also illustrates how to accomplish several things with the Windows® O/S using C.

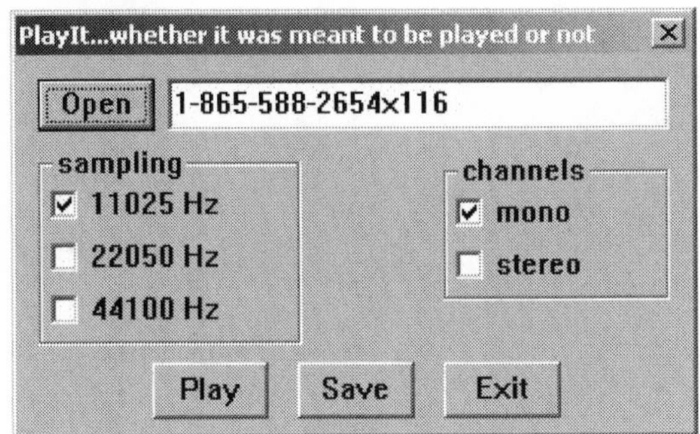

Appendix E: Triangle Gridder

This doesn't have anything to do with orthogonal functions, but it does illustrate some code that we used in Chapter 7 to determine the convex bounding polygon. In the folder examples\gridder you will find a little Windows® program that creates randomly shaped domains and then breaks them into triangular elements. It runs until you press Esc or alt-F4. A typical grid is shown below:

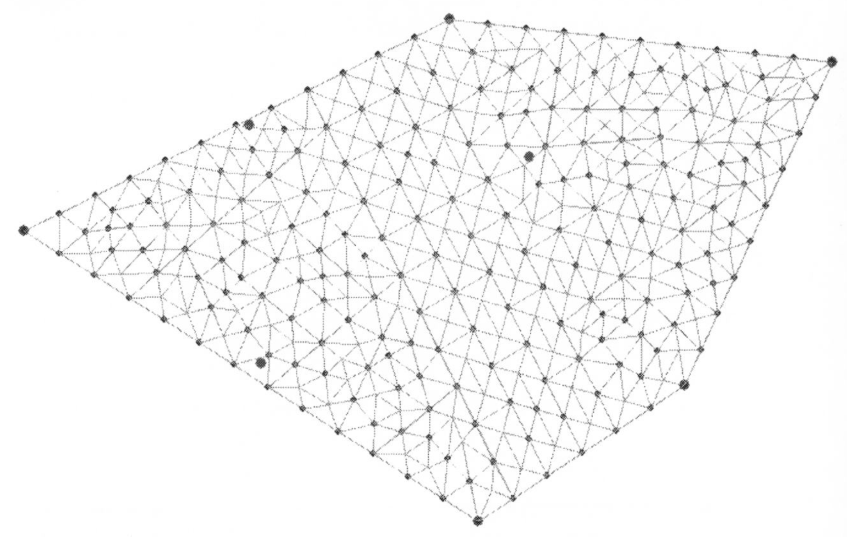

The algorithm tries to make equilateral triangles instead of acute wedges that are computationally problematic.

also by D. James Benton

3D Rendering in Windows: How to display three-dimensional objects in Windows with and without OpenGL, ISBN-9781520339610, Amazon, 2016.

Curve-Fitting: The Science and Art of Approximation, ISBN-9781520339542, Amazon, 2016.

Differential Equations: Numerical Methods for Solving, ISBN-9781983004162, Amazon, 2018.

Evaporative Cooling: The Science of Beating the Heat, ISBN-9781520913346, Amazon, 2017.

Heat Exchangers: Performance Prediction & Evaluation, ISBN-9781973589327, Amazon, 2017.

Jamie2 2nd Ed.: Innocence is easily lost and cannot be restored, ISBN-9781520339375, Amazon, 2016-18.

Little Star 2nd Ed.: God doesn't do things the way we expect Him to. He's better than that! ISBN-9781520338903, Amazon, 2015-17.

Living Math: Seeing mathematics in every day life (and appreciating it more too), ISBN-9781520336992, Amazon, 2016.

Lost Cause: If only history could be changed…, ISBN-9781521173770. Amazon 2017.

Mill Town Destiny: The Hand of Providence brought them together to rescue the mill, the town, and each other, ISBN-9781520864679, Amazon, 2017.

Monte Carlo Simulation: The Art of Random Process Characterization, ISBN-9781980577874, Amazon, 2018.

Nonlinear Equatiions: Numerical Methods for Solving, ISBN-9781717767318, Amazon, 2018;

Numerical Calculus: Differentiation and Integration, ISBN-9781980680901, Amazon, 2018.

ROFL: Rolling on the Floor Laughing, ISBN-9781973300007, Amazon, 2017.

A Synergy of Short Stories: The whole may be greater than the sum of the parts, ISBN-9781520340319, Amazon, 2016.

Thermodynamics - Theory & Practice: The science of energy and power, ISBN-9781520339795, Amazon, 2016.

Version-Independent Programming: Code Development Guidelines for the Windows® Operating System, ISBN-9781520339146, Amazon, 2016.

Made in the USA
Middletown, DE
04 October 2021